HOW TO REPAIR

HOME LAUNDRY APPLIANCES

2ND EDITION

HOW TO REPAIR

HOME LAUNDRY APPLIANCES

2ND EDITION

BEN GADDIS

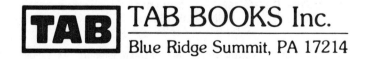

TAB BOOKS Inc.
Blue Ridge Summit, PA 17214

SECOND EDITION

FIRST PRINTING

Library of Congress Cataloging in Publication Data

Gaddis, Ben W.
How to repair home laundry appliances.

Includes index.
1. Washing-machines—Maintenance and repair.
2. Clothes dryers—Maintenance and repair.   I. Title.
TT997.G27   1986   648'.1'028   85-27838
             ISBN 0-8306-9562-1
             ISBN 0-8306-0662-9 (pbk.)

Cover photographs courtesy of Lowe's of Chambersburg, PA 17201.

Contents

HOW TO REPAIR

HOME LAUNDRY APPLIANCES

2ND EDITION

Introduction

THERE IS NOTHING PARTICULARLY FRIGHTENING ABOUT REpairing a washer or dryer, if you know what you are looking for and where to look for it. These appliances contain ordinary mechanical and electrical components that perform simple repetitive tasks in response to timed commands. To repair the appliances it is necessary to know four key facts about their operation:

- What task is not being performed as it should.
- When should that task be performed.
- Which device is responsible for that task.
- How that device is supposed to perform the task.

Armed with these facts, it is a matter of observing the cyclic operation of the machine to determine where the failure is occurring. The purpose of this book is to aid you in understanding troubleshooting techniques.

If you have a background in basic electricity or electronics, this book will prove to be the only tutor you need to actually set up shop to go into the home laundry-appliance service business on a full-time or part-time basis. If you are a stranger to the electron, you'll find enough information in this book to help you through any washer or dryer problems you're ever apt to face. You have to learn a few simple rules: what you can and can't do; how to troubleshoot using progressive and logical procedures; and how to spot basic appliance problems using your own natural senses. Before long you'll understand the operational differences between your machine's timed functions, and you'll have a "handle" on most laundry appliance problems.

If you are already doing washer and dryer repair as a business, you will find the book helpful in expanding your present knowledge and developing better troubleshooting skills. If you are a self-styled handyman and enjoy doing your own repair work, you will appreciate the simplified explanations of circuit operations and the troubleshooting tips that accompany them.

Some technical manuals of this type discuss specific makes and models of machines, thus limiting their application. Others are loaded with trouble-specific remedy tables (again making them of limited value). These approaches have been avoided in this book. Instead the emphasis is placed on describing the basic cycles used by all laundry machines, and explaining various techniques to be used to isolate failures of specific components.

Overview

T HE MODERN HOME LAUNDRY IS MORE THAN JUST A CON-
venience. Any housewife will tell you that her washer and
dryer are her most prized appliances. This should be no surprise.
Modern washers and dryers with cycles for any fabric, along with
today's knit and permanent-press fabrics, eliminate much of the
washday drudgery of ten years ago. And ironing, once a time-
consuming and difficult chore, has been greatly simplified. In view
of this it is easy to understand why most housewives are upset when
they have to call the appliance serviceman. What could be more
frustrating than a waist-high stack of dirty laundry on Monday
morning with the washer gone mad and the laundry room ankle
deep in sudsy water? Or when, on the coldest, wettest day of the
year and with two tubs of wet clothing, the dryer tumbles just fine
but doesn't get hot enough to melt an ice cube?

Then comes the worry about how long repairs will take and
how much they will cost. At today's prices these worries are un-
derstandable. Increases in the cost of parts, labor, and general operation have caused the cost of appliance repair to climb along with
prices of other goods and services. In some areas of the country
parts are often difficult to obtain. This can have a great bearing
on the time required to make repairs.

Considering experiences such as these, the irritation felt by the
owner of a defective appliance seems only natural. Keep this in mind

during your contacts with customers. How you handle the first few moments of the service call can have a great effect on the customer's attitude towards you and your services. Don't make promises that you can't keep, but be as reassuring and as helpful as you can. This will relieve much of the tension sometimes experienced between the customer and the serviceman.

Let's start with the first contact—usually a phone call. Just how you answer the call can have considerable effect on the customer. Assume a friendly, cheerful voice to determine the nature of the complaint. Find out as much as you can about the equipment: make, model, and the approximate date of purchase. This information will help you determine what parts and special tools should be taken on the service call. Establish with the customer a definite time you will arrive. Don't add to the aggravation by keeping the customer waiting all morning when you could possibly establish a time to within an hour.

A well-equipped service truck represents a large investment, and the selection of parts to have available in the truck is sometimes difficult to determine. This is extremely important, however. Not only will a well-stocked service truck save you time and provide a better profit margin, but it means that appliances will not be out of service for too long. Experience shows that most customers don't mind the repair bill as much as they mind the appliance being out of service while you wait for parts that must be ordered. The selection of parts you carry in the service truck and stock at the shop will depend on several factors. Perhaps the most obvious factor is the make of appliances you normally service. The decision must ultimately be made from your own experience.

Your conduct, attitude, and appearance will make an immediate impression on your customer. If you are polite and helpful with a professional attitude and professional equipment, you are off to a good start. A sloppy appearance or a casual attitude can destroy the customer's confidence in your ability to service complex equipment.

A profitable home laundry service must be more than just a fix-it business. These machines represent a considerable investment by the consumer, and he rightly expects high-quality service for the price he must pay. This book can help you render that professional service.

HOME LAUNDRY CENTER

Modern home laundries include automatic washers and dryers and are a very important part of today's household. Most homes today are built with a specific area set aside for the laundry. Adequate water, drainage, electrical power, and a gas supply are provided for these areas.

Most modern washers are completely automatic and include special cycles for delicate and permanent-press fabrics. A presoak cycle and a lint filter are also featured. A water hookup behind the washer furnishes hot and cold water. A temperature control blends hot and cold water to achieve the desired temperature. A level control fills the washer with the correct amount of water for the load to be washed. A proper electrical service of at least 15 A is available behind the washer. Although not part of the washer itself, a level floor provides proper support for convenience and safety.

A 240 V outlet behind the electric dryer supplies power for heat and automatic operation. A flexible vent pipe connects the dryer exhaust to a vent installed through the wall. This expels heat and lint fibers from the machine to the outside atmosphere. In addition to the requirements of the machines themselves most installations include storage space for soiled clothing, laundry supplies, and ironing equipment.

Many laundries in older homes do not feature all the conveniences discussed here. Those that do may have conveniences that are improperly installed. Poor plumbing and venting and dangerously inadequate wiring are sometimes encountered. For this reason we will discuss the wiring needs of modern home laundries. This is not to teach you to wire the laundry but to recognize proper and improper wiring. The appliance serviceman is frequently the first professional familiar with electrical equipment with whom the homeowner discusses his laundry wiring problems.

One word of caution: Most states, counties, and cities prohibit any additions, alterations, or modifications to home wiring by anyone other than a licensed electrician. You are cautioned against offering such service without first consulting your local government offices. The city (or county) engineer can give you the rules for your locality.

ELECTRICITY

Although some of the electricity we use is generated locally, most of it comes to us through a very complex distribution system. Even electricity generated for special purposes must be routed through some sort of distribution system. Let's take a look at the system that brings electricity into the home.

Generation

Electricity in the United States has been generally standardized—the electricity used in Boston has the same characteristics as that used in Omaha or Seattle. Two values of voltage are usually supplied to modern residences—120 V and 240 V. The voltage may vary slightly about these values, depending on the particular distribution system, but never more than a few volts. Frequency is set at 60 hertz (Hz)—60 complete cycles each second. This voltage may be generated in several ways. Some cities have their own generating plants while others, especially smaller towns, bring power in from some remote location. The generator may be hydroelectric or steam turbine, depending on the resources available.

Large hydroelectric dams are located in many areas of the United States where there is an abundance of rivers and streams. The Northwestern United States and the Central States east of the Mississippi River are prime examples. Generators in these dams convert mechanical motion to electricity. The dam itself blocks the natural flow of the stream and causes the water to back up, forming a large man-made lake. Near the dam this water is quite deep an exerts an extremely high pressure on anything near the bottom. That pressure with the natural downward flow of the water, turns the generators.

Steam turbine generators are similar in some respects to hydroelectric generators. The main differences are in the turbine itself and its drive mechanism. Boilers are used to generate the steam that spins the generator. Fuel for the boiler will vary with geographic area.

Conductors

The electric power is brought to the home from the generators by way of conductors. Conductors vary in size, shape, insulation, and other qualities, depending on their intended use. Copper and aluminum have become the most common metals used as conductors. Aluminum has the advantages of weighing and costing less than copper. It also has the disadvantage of lower current-carrying capabilities, requiring a larger conductor to carry the same amount of current.

Regardless of the type of material used in making the wire, all conductors have certain common properties with which you should be familiar. All material has a specific resistance, measured in ohms and symbolized by the Greek letter omega (Ω). Resistance is different for every material and is usually quoted as the amount of resistance present in a circular-mil foot of the material. An example of the circular-mil foot is shown in Fig. 1-1, the conductor being one mil in diameter and one foot in length.

By using the specific resistance per foot, the resistance of any length or diameter of a material can be computed as

$$R = \frac{\varrho L}{A}$$

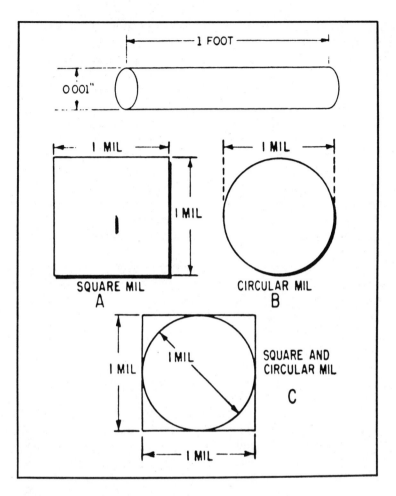

Fig. 1-1. The circular mil-foot.

where

R = resistance in ohms

ϱ = specific resistance in ohms per circular-mil foot of the material

L = length of the conductor in feet

A = cross sectional area in circular mils

From this formula you can see that as the length of a conductor is increased, the resistance is increased; and if the cross sectional area of the conductor is increased, the resistance is decreased. Resistance then varies inversely with the cross sectional area and inversely with the square of the diameter (since area = D^2). Therefore, if the diameter is halved, the area is quartered and resistance is quadrupled. If the diameter is doubled, the area is quadrupled and resistance is quartered.

The formula shows that a small wire will offer more resistance to current than a large wire of the same material. Since power is computed as $P = I^2 R$ this increase in resistance causes the smaller wire to consume more power than a larger wire carrying the same current. This power is dissipated as heat. Should the current become high enough the heat will become so intense that the insulation and the wire will melt. This is why large conductors must be used to carry high currents.

The loss of power in a conductor also means a loss of voltage (E) at the load. Assume a circuit (Fig. 1-2A) that has a 110 V generator supplying power to a 100Ω load through two wires having 5Ω resistance each. The generator sees a total resistance (R) of 110Ω, the resistance of each wire *plus* the load resistance. Under these conditions a total current (I) of 1 A is present in the circuit: $I = E/R = 110 V/110Ω = 1 A$.

With 1 A of current the three resistances will cause voltage drops around the circuit as shown in Fig. 1-2B. As can be seen, the voltage across the load will be 100 V instead of 110 V. Although this particular example is somewhat exaggerated, it illustrates the results of using a conductor that is too small in diameter or too long in length. This condition causes the load (appliance in our case) to operate improperly—underheating elements, overheating or burning motors, etc. From this example we see the importance of using the proper wire for the job.

Wire size has been standardized throughout the United States. The size is given as a number; the smaller the number the larger the size. Common sizes run from 0000, which has a cross-sectional area of 0.166 square inches down to 40 gage, having an area of 0.0000078 inch. A conductor may be one solid wire or a group of smaller wires twisted together. The twisted strands are used where a large conductor with good flexibility is needed.

In addition to wire size, the insulated covering is also important. The conductor may be a single insulated wire or many conductors in one cable, each separated from the other by its own insulated covering. In some cases the cable is covered with metal braid (armor) to protect the insulation. Other installations require the conductors to be run through a metal pipe called conduit.

Most electrical distribution systems make use of a wiring code in which the insulated covering of each conductor is colored according to its use. This is especially true in residential and industrial applications.

Neighborhood Distribution

Electrical power is brought into the home from the generating source by high-tension wires. At the generating station the voltage output is typically 10,000—20,000 V. In order to avoid current

*Fig. 1-2. Voltage loss in an electri-
cal conductor.*

losses, step-up transformers increase the voltage to somewhere between 20,000 and 150,000 V. By transmitting the power at a very
high voltage and minimum current power, losses in the wire are
reduced. This allows the use of smaller wire.

From the step-up transformers the electrical power is carried
to local substations where the voltage is reduced and routed to the
surrounding neighborhoods. In some areas this is done by the familiar utility poles while other areas use an underground transmission system. Located throughout the substation area are
transformers that step down the voltage to a level where it can be
used in the home. You have probably seen these transformers at
the tops of utility poles in some areas.

Figure 1-3 illustrates the function of the utility pole transformer.
The power from the substation is routed to the local step-down
transformers, where the high voltage is reduced to two values—
120 V and 240 V. Some older homes have only two-wire service,
the neutral and one leg of the secondary. Newer homes have three-
wire service, the neutral and both legs of the secondary. The voltage between the neutral and any one leg will be about 120 V, while
240 V is present between the two "hot" legs. Since the National
Electric Code (NEC) requirements have brought changes in service
entrance requirements, we concentrate our efforts on the 240 V
installation.

7

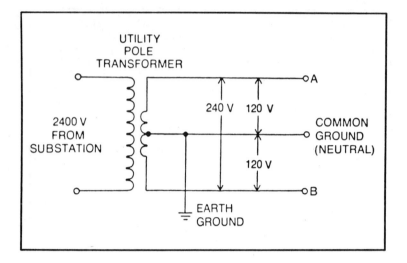

Service Entrance

The power is brought from the pole to the service entrance meter. The meter is equipped with contacts that plug into a socket similar to that shown in Fig. 1-4. Three conductors are used in a typical 240 V installation. Between the red and black, leads coming from the pole will be approximately 240 V. The voltage between either the black or red and the neutral will be about 120 V. The contacts on the meter connect the pole leads to leads entering the circuit-breaker box. All current used is monitored by the meter. Notice that the neutral is grounded to earth by a grounding conductor.

Heavy-gage conductors connect the meter socket to the circuit-breaker box. This box acts as the distribution point from which all circuits are routed. In some older installations, fuses will be used instead of the more convenient circuit breakers. The fuse must be replaced if a circuit is overloaded. If an overload occurs in a circuit protected by a circuit breaker, the breaker need only be reset once the overload condition has been corrected.

If you traced the electrical circuit in a modern home from the point where it enters the home, immediately after the meter you would come to the distribution box or breaker box. This is the point from where each circuit is routed to its area of use. The NEC has specific rules which must be followed when these circuits are installed. Although you will not likely make any changes or additions to these circuits (in most cases the law forbids changes or additions) you should, for safety, be familiar with code requirements. The heavy loads placed on the electrical system by home laundry equipment make it necessary for the appliance technician to know enough about home wiring systems to recognize inadequate or unsafe wiring. The breaker box is one of the most important parts of the home

Fig. 1-3. The neighborhood electrical distribution system uses an underground transmission system or utility poles.

wiring system. Before examining the breaker box itself let's look briefly at the general requirements placed on the home wiring system by the NEC.

The lighting circuit must provide a minimum of three watts of lighting for each square foot of living space. That's just for lighting, and the bare minimum. Appliance requirements are in addition to the lighting load. Two individual 20 A, 1500 W appliance circuits must be provided. These circuits are separate from, and in addition to, any circuit for a special appliance, such as an electric range or water heater. You may not find all these features in many older homes, but any new construction must comply with these requirements.

Service entrances, and therefore breaker boxes, are identified by the maximum amount of current to be supplied through that entrance. In the early days of home electrical service, installations were as small as 30 A. The 60 A and 100 A services are more typical today. Some entrances are capable of handling 160 A. The diagram in Fig. 1-5 depicts a 100 A service-entrance breaker panel. The neutral lead is brought in and tied to the neutral tie for all circuits. Notice that no fuse or breaker is found in the neutral leg of

Fig. 1-4. The meter hook up includes contacts that plug into a socket.

FROM METER

A B

MAIN BREAKERS
100A EACH

35 A
BREAKERS

35 A BREAKERS

TO
RANGE

TO
DRYER

NEUTRALS

20 A

15 A

20 A

15 A

TO
WATER
HEATER

120 V
BUS

120 V
BUS

TO
WATER
HEATER

NEUTRAL
BUS
BAR

any circuit. This is required by the NEC. The neutral must be un-broken throughout the entire wiring system.

The hot wires (red and black) run from the meter socket to the main circuit breaker or fuse. Although separate 100 A fuses or circuit breakers are provided for each leg, they are mechanically connected (ganged) so that if one circuit is opened by tripping the breaker (or breaking the fuse) the other circuit will also be interrupted. Each of these 100 A breakers supplies its own bus, from which the various branch circuits emerge. Notice that between any hot leg and neutral there is 120 V, while between the two hot legs there is 240 V. For any one of the lighting circuits it will be necessary to run one neutral wire and one hot wire. The neutral ties directly to the neutral bus while the hot wire is connected to one of the hot buses through a circuit breaker.

In addition to its separate 120 V bus, each 100 A breaker supplies one leg of the 240 V circuits for the electric range and electric dryer. Both these appliances are protected by ganged 35 A breakers in each 120 V leg. Similar to the main breaker, these spe-

Fig. 1-5. Diagram of a 100A service-entrance breaker panel. Notice that no overload devices are found in the neutral lines.

cial breakers open both legs to the appliance simultaneously. Again, an unbroken neutral is supplied from the neutral bus. The water heater receives its power through two separate 15 A breakers.

Each of the lighting and convenience circuits are protected by a 15 A circuit breaker. These circuits supply electrical power for room lighting and such items as TVs, radios, and other devices. Each of these circuits should be limited to no more than ten outlets, but in no case should the load be more than 15 A.

The NEC requires that all new installations include two separate 20 A appliance circuits. These circuits are not to be used for any purpose other than appliance operation. In our example in Fig. 1-5, the top breakers on each bus provide these circuits.

One important aspect of using the distribution box is balancing the load. Circuits should be connected across the supply buses so that they balance or equalize the load on each of the 120 V hot buses as much as possible. The importance of this is illustrated in Fig. 1-6.

Assume that the windings in Fig. 1-6A represent the secondary of a power transformer whose output is 240 V between L1 and L2, 120 V between N and L1, and 120 V between N and L2.

Fig. 1-6. Balancing the load.

An ammeter has been inserted in each leg for illustration. Now, if a load having 12Ω is placed across the 120 V circuit between N and L1 the current will be 10 A: $I = E/R = 120 V/12Ω = 10 A$.

Under these conditions with no load, and therefore no current flow across N to L2, ammeter A1 will read 10 A; A2, 10 A; and A3, 0A. This means that if we ignore the power losses in the wire itself 10 A flows through both the L1 and N conductors.

Now let's add a second 12Ω load, but add this new resistance between N and L2 as shown in Fig. 1-6B. This load would also draw 10 A as indicated by A3. The effect of the current in the N leg is what we are interested in here. Notice that with the two loads connected as shown we have a 24Ω circuit connected across 240 V between L1 and L2. Since these loads are equal each must drop half the total voltage. This means that half the voltage is dropped across the load between N and L1, and the other half is dropped across the load between N and L2. Theoretically, we see that the voltage at the junction of the two loads (the N conductor) will be zero and there would be no current flow through N. This assumption illustrates how a balanced load can reduce current flow in the neutral conductor. Since any conductor has some resistance, current through that resistance causes voltage drops along the line. If current in the neutral conductor is reduced by balancing the load, line losses can be kept to a minimum.

Most contractors are careful to balance loads in new electrical installations. You may encounter some systems where homeowners have done their own add-on work to the original circuits, however. These may or may not be according to the NEC and they may or

Fig. 1-7. An add-on installation connecting a 60A service. The 60A main breaker must be changed to 100A.

may not be balanced. You should be extra careful when working with such circuits since you have no assurances that proper safety precautions have been taken.

In many cases older homes have been remodeled and, to save on remodeling costs, the existing wiring was retained. The extra demand placed on the existing service, however, requires the addition of new circuits. In some cases the breaker box has been replaced while in others the extra power is provided from a small add-on unit. Figure 1-7 illustrates how to install such add-ons. The installation shown was originally a 60 A service. Note the similarities between the 60 A breaker box in Fig. 1-7 and the 100 A box shown in Fig. 1-5.

A heavy conductor is connected between the neutral bus in the existing box and the new box. In this particular case the 35 A range circuit was unused. Therefore, the additional breaker box can be added by connecting it to the range breaker. Two 15 A circuits can be added to each hot leg of the 240 V range circuit.

General Home Wiring

The importance of proper wiring was mentioned earlier. Wiring too small in diameter or too long causes excessive line losses. At best, incorrect wiring can cause inefficient operation of appliances; at worst, shocks and fire hazards. From the efficiency standpoint alone you should be concerned about wiring that supplies power to appliances you must service. There are two different home wiring systems you may encounter in your service work.

You will find the two-wire system in older homes and in some illegal do-it-yourself modifications. Installation of this system is no longer allowed under NEC rules. Figure 1-8 shows how a two-wire circuit is run from a circuit-breaker box to a bedroom.

Notice that one wire is brought from the neutral bus to a junction box where the lighting and convenience outlets in this branch are tied together. By code requirements the insulating wrapper of this neutral conductor must be white. From the connection box the white neutral wire is supplied directly to the lighting fixture, and one white wire ties all the convenience outlets together in parallel.

The hot wire is run from the circuit breaker to the junction box and parallels the convenience outlets. Notice that the lighting circuit is different, however. Instead of going directly to the lighting fixture, as did the neutral, the hot wire goes first through a switch then to the lighting fixture. The NEC requires that the hot wire be covered with a black insulative wrapper. These black and white color codings must be maintained throughout the system.

Although a switch in the neutral line could be made to control the lighting fixture, the code requires that no switches, breakers, or any other device capable of interrupting the neutral circuit be

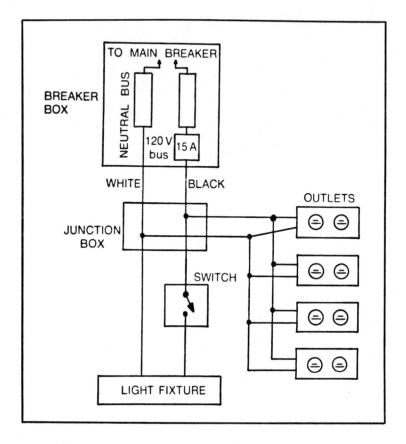

installed. Notice in Fig. 1-8 that without any breakers or switches the neutral terminal on the lighting fixture is common (zero resistance) with the neutral terminals of the convenience outlets. Also, all the neutral terminals of the wiring system are common since all neutrals are uninterrupted and tie to the neutral terminal block or bus in the circuit-breaker panel.

If you traced the white neutral back to the service entrance you would find that this wiring is grounded through the service-entrance ground. Here a long metal stake or plate is sunk deep into the earth, and the neutral leg is tied to that grounding wire.

The quickest and most obvious way to recognize the two-wire circuit is through examination of the convenience outlets. Figure 1-9 illustrates the common two-wire convenience outlet. One black and one white wire bring the neutral and hot legs to opposite sides of the outlet. There may be more than two wires in the box, however. These wires would parallel the remaining outlets on that branch. Notice that the metal box housing the outlet is not connected in any way to the electrical system.

A variation of the two-wire system can be identified by the type of two-wire outlet used. More recent two-wire installations use a

Fig. 1-8. Typical two-wire branch circuit running from a breaker box to the bedroom. The NEC requires all "hot" wires be covered with black insulation. The neutral wire is covered with white insulation.

polarized outlet. These outlets are similar to the two-wire units just discussed except for the physical size of one of the connectors in the outlet. Close examination of the polarized outlet will reveal that one of the connectors is slightly larger than the other. When polarized outlets of this type are used the neutral wire should be connected to the side of the outlet that is common to the larger connector.

The polarized outlet came into being with appliances that have polarized power cords. These cords have one prong on the plug larger than the other. Most of their applications have to do with electronic equipment, such as radios and TVs, and have little to do with our subject. But you should be able to recognize these installations when you see them. Keep the following in mind:

- Under no circumstance should polarized connectors or outlets be modified to defeat the polarization feature. This can result in severe shock hazards to those who operate appliances connected to these circuits.
- Never assume that polarized outlets are wired properly. Test the circuit to insure that the neutral leg is properly wired.
- Never trust wiring color codes. In spite of NEC rules, you may encounter systems where safety precautions have been violated or ignored.

The modern method of home wiring, and the only one permitted by the present code in new installations, is called the three-wire system.

Fig. 1-9. The two-wire convenience outlet.

Wiring for this system uses the black and white color-code system, as used in the two-wire system, plus an additional wire. The third wire is not insulated and is called the *bare ground*. Outlets for these systems have three connections: the hot wire, the neutral, and a ground (Fig. 1-10). This bare wire is tied to the ground connector and the metal boxes throughout the system. Code requirements state that this ground, similar to the neutral, must be unbroken throughout the entire system.

The ground wire is sometimes referred to as the *nonconducting* neutral since under normal operating conditions no current flows in this wire. Notice that nowhere in the system is the ground connected in any way to either the neutral or hot sides of the circuits. Instead, somewhere near the service entrance the ground wire is connected to a cold-water pipe or a grounding rod that is separated from the grounded neutral. Hot water, gas, or plastic pipes cannot be used for the grounding device. In some areas where plastic pipe is used as the cold-water line, a grounding rod similar to that employed with the neutral must be used. This is common in rural areas where plastic pipe is used between the well and the cold-water entrance to the house.

Grounding the system offers the advantage of safety over the common two-wire system. One important safety feature is illustrated in Fig. 1-11. A motor-driven appliance with a metal cabinet is shown connected to a two-wire 120 V system by a two-wire power cord and a two-prong outlet. Notice how the neutral wire and the hot wire connect to the breaker panel. As long as the insulation and electrical connections in the appliance remain in good shape there is no problem.

Fig. 1-10. The three-wire grounded outlet.

NEUTRAL

HOT (NARROWER SLOT)

NEUTRAL
TERMINALS

HOT TERMINALS
ON THIS SIDE

GROUND
TERMINAL

GROUND CONNECTION

But what happens if one of the hot wires becomes disconnected or the insulation breaks down and the hot side of the line becomes tied to the metal cabinet? A dangerous situation evolves but no permanent damage is sustained.

Everything remains normal and no external symptoms of trouble appear until some housewife attempts to use this appliance. If she is standing on an uninsulated floor or is accidently touching another metal object, there is an electrical path from the 120 V, 20 A breaker through her body to the object she is touching. Severe electrical shock or even death could be the result. Had this appliance been properly connected to a three-wire circuit, the possibility of this accident ever occurring may not have been completely eliminated, but the degree of danger would have been lowered.

Look at Fig. 1-12. Same appliance, same defective hot lead, only this time the cabinet is connected to a three-prong ground outlet. The ground connector is then tied to the system ground by the bare conductor explained with Fig. 1-9. Under these conditions current can flow from the 20 A breaker through the faulty connection to the cabinet of the appliance. Since the cabinet is connected to ground through the three-wire system, current will start to flow before anyone or anything comes in contact with the cabinet. Should this current become heavy enough, such as cases involving direct shorts, the circuit breaker will trip and no power will be present at the outlet. Although this doesn't entirely remove the possibility of electrical shock from a defective appliance, it does decrease the danger significantly.

The need for adequate grounding of the home electrical system cannot be overemphasized. The three-wire scheme provides for protection if the wiring rules of the NEC are followed. As an added safety, however, it is good practice to provide an additional ground for each major stationary appliance. The ground applied to the third wire of the system is called the system ground to distinguish it from the equipment (individual) ground. The code requires that the resistance between the system ground and equipment ground be less than 25Ω.

Individual equipment grounds can be attached from metal parts of the cabinet to a good ground such as a cold-water pipe. This added protection should be provided when such appliances are installed in damp spaces, such as basements, or when there is any doubt as to the quality of the system ground.

Wire Types and Sizes

Types and sizes of wire used in home wiring should be recognized by any technician working with major electrical appliances. You should know enough in this area to advise your customers concerning the wiring of their homes, especially the appliance circuits.

Three types of wire are used in home wiring and they are illustrated in Fig. 1-13. *Conduit* wiring in Fig. 1-13(A) consists of thin pipe through which the wire is run. The conduit protects the insulative wrapping of the conductors against moisture and abrasion. There are specific limits as to the size and number of conductors that may be carried in any particular size of conduit. These details are not required for our purposes here, however. Properly installed conduit wiring does not require a separate ground since the metal piping should be well grounded.

Similar to, and sometimes confused with, conduit is *metallic shielded cable* as in Fig. 1-13(B). This offers many of the advantages of conduit and is somewhat easier to run. It is available in several sizes coded by number and size of conductors. The metallic shielding of this cable can be used for a grounding conductor.

One of the most popular types of wiring in use today is the *nonmetallic sheathed cable* shown in Fig. 1-13(C). Nonmetallic cable is

Fig. 1-12. The three wire grounded appliance outlet provides better safety against potential hazards.

Fig. 1-13. The general types of electrical cables include conduit, metallic shielded and nonmetallic sheathed cables.

more flexible and much lighter than sheathed cable or conduit. Depending on where it is to be installed, this cable may be weatherproofed for outdoor, underground, or indoor use. There are also types for use in hot, humid, or extremely dry areas.

This cable also comes in different sizes coded by the number of conductors and the wire size. A group of letters is used to designate the type of cable according to the area of intended use. For example, TW is thermoplastic cable for general use in wet areas. Size is designated by a combination of numbers. The number 10-2, for example, is used to designate a cable containing two conductors of 10-gage wire. Three 12-gage conductors contained in a single sheath would be designated 12-3. Nonmetallic cable comes with or without the separate ground conductor. The ground conductor does not have an insulative wrapper.

Wire size is as important as the insulative qualities of the wrapper. Common American wire sizes are designated by numbers—the larger the number, the smaller the diameter of the wire. The smallest wire size allowed in home wiring is 14-gage. Figure 1-14 is a table of common wire sizes as determined by the American Wire Gage (AWG) standard accepted by the NEC. Figure 1-15 is a typical gage used to measure the diameter of wires.

The length and cross-sectional area must also be considered when selecting the proper wire size. The cross-sectional area (gage) must be increased if the length of the run is increased and the current is held constant. This is because of the increased voltage drop across the line.

Appliance Circuits

The requirements placed on lighting and convenience circuits apply to appliance circuits as well. In addition, there are several more requirements. First let's consider the 240 V appliance circuits, in particular the water heater and electric dryer circuits. You may recall from the discussion of the service entrance that both

Gage number	Diameter (mils)	Cross section		Ohms per 1000 ft.		Ohms per mile 25° C. (= 77° F.)	Pounds per 1000 ft.
		Circular mils	Square inches	25° C. (= 77° F.)	65° C. (= 149° F.)		
0000	460.0	212,000.0	0.166	0.0500	0.0577	0.264	641.0
000	410.0	168,000.0	.132	.0630	.0727	.333	508.0
00	365.0	133,000.0	.105	.0795	.0917	.420	403.0
0	325.0	106,000.0	.0829	.100	.116	.528	319.0
1	289.0	83,700.0	.0657	.126	.146	.665	253.0
2	258.0	66,400.0	.0521	.159	.184	.839	201.0
3	229.0	52,600.0	.0413	.201	.232	1.061	159.0
4	204.0	41,700.0	.0328	.253	.292	1.335	126.0
5	182.0	33,100.0	.0260	.319	.369	1.685	100.0
6	162.0	26,300.0	.0206	.403	.465	2.13	79.5
7	144.0	20,800.0	.0164	.508	.586	2.68	63.0
8	128.0	16,500.0	.0130	.641	.739	3.38	50.0
9	114.0	13,100.0	.0103	.808	.932	4.27	39.6
10	102.0	10,400.0	.00815	1.02	1.18	5.38	31.4
11	91.0	8,230.0	.00647	1.28	1.48	6.75	24.9
12	81.0	6,530.0	.00513	1.62	1.87	8.55	19.8
13	72.0	5,180.0	.00407	2.04	2.36	10.77	15.7
14	64.0	4,110.0	.00323	2.58	2.97	13.62	12.4
15	57.0	3,260.0	.00256	3.25	3.75	17.16	9.86
16	51.0	2,580.0	.00203	4.09	4.73	21.6	7.82
17	45.0	2,050.0	.00161	5.16	5.96	27.2	6.20
18	40.0	1,620.0	.00128	6.51	7.51	34.4	4.92
19	36.0	1,290.0	.00101	8.21	9.48	43.3	3.90
20	32.0	1,020.0	.000802	10.4	11.9	54.9	3.09
21	28.5	810.0	.000636	13.1	15.1	69.1	2.45
22	25.3	642.0	.000505	16.5	19.0	87.1	19.4
23	22.6	509.0	.000400	20.8	24.0	109.8	1.54
24	20.1	404.0	.000317	26.2	30.2	138.3	1.22
25	17.9	320.0	.000252	33.0	38.1	174.1	0.970
26	15.9	254.0	.000200	41.6	48.0	220.0	0.769
27	14.2	202.0	.000158	52.5	60.6	277.0	0.610
28	12.6	160.0	.000126	66.2	76.4	350.0	0.484
29	11.3	127.0	.0000995	83.4	96.3	440.0	0.384
30	10.0	101.0	.0000789	105.0	121.0	554.0	0.304
31	8.9	79.7	.0000626	133.0	153.0	702.0	0.241
32	8.0	63.2	.0000496	167.0	193.0	882.0	0.191
33	7.1	50.1	.0000394	211.0	243.0	1,114.0	0.152
34	6.3	39.8	.0000312	266.0	307.0	1,404.0	0.120
35	5.6	31.5	.0000248	335.0	387.0	1,769.0	0.954
36	5.0	25.0	.0000196	423.0	488.0	2,230.0	0.0757
37	4.5	19.8	.0000156	533.0	616.0	2,810.0	0.0600
38	4.0	15.7	.0000123	673.0	776.0	3,550.0	0.0476
39	3.5	12.5	.0000098	848.0	979.0	4,480.0	0.0377
40	3.1	9.9	.00000078	1,070.0	1,230.0	5,650.0	0.0299

the electric range and electric dryer have their own circuits, complete with a fuse or circuit breaker. The water heater has its own separate branch taken from the 120 V buses. Under no conditions should any other circuit be tied into these circuits, not only for the obvious reason that they are 240 V circuits but also because the water heater circuit is not supplied with a neutral wire.

Automatic washers and gas dryers require 120 V for operation. These should be at least 20 A branch circuits. In addition to having three-prong grounded outlets it is good practice that these appliances be equipped with an equipment ground. No effort should be spared to make the laundry equipment as safe as possible. The inherent danger of water and electricity in such proximity requires the service technician to insure safe operation by the customer. When servicing laundry appliances always check the grounding system and general condition of the wiring as a part of your service call.

ELECTRICAL SAFETY

Many electrical appliance operators are not aware of the safety precautions that should accompany their operation. You do your customers a great disservice when you do not instruct them in the safe operation of their appliances. When we study the installation of these appliances, we will see the details of safe home laundries. For now, let's look at some of the safety devices in the appliance branch circuits. Proper grounding has already been discussed and will be brought to mind again in later chapters.

Fuses

You will encounter different types of fuses (see Fig. 1-16) in

Fig. 1-15. A typical gage used to determine the diameter of a wire.

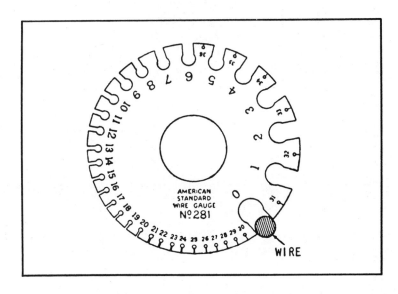

AMERICAN
STANDARD
WIRE GAUGE
N⁰ 281

WIRE

home wiring systems. The most common is a plug fuse. These fuses range in link size up to 30 A. The threaded base of these fuses is the same size as an ordinary incandescent lamp. As seen in the illustration, there are three basic parts to the fuse: the shell screw, the centerpost, and the fusible link. The fusible link connects the centerpost to the shell screw. When a load draws current higher than the rating of the fuse the heat generated by the current will melt the fusible link.

These fuses have several disadvantages, not the least of which is the danger of overfusing. Frequently, when a customer experiences repetitive blown fuses on a particular branch circuit, he will overfuse (use a larger capacity fuse) in the fuse holder. Occasionally you may find the fuse defeated by a conductor, such as a penny, inserted behind the fuse. Overfusing is a very dangerous practice and should be pointed out to the customer whenever encountered.

There are times when it is desirable, however, to allow brief high currents in certain branch circuits. For example, circuits supplying electric power to motors are subject to momentary high currents needed to start the motor. These currents would blow the ordinary fuse. To prevent this, time-lag or slow-blow fuses (shown in Fig. 1-17) are used. Instead of melting the fusible link, a constant overload causes the solder to melt and release the spring which opens the circuit. Motor currents do not last long enough to melt the solder. Therefore, these fuses can handle short periods of high current required in motor circuits.

To prevent overfusing, a tamper-resistant plug fuse is available. It is constructed in two separate parts as shown in Fig. 1-18. The adapter, similar to the plug, comes in two sizes. One size is for fuses up to 15 A, the other for fuses of 16—30 A. Due to the construction of the adapter ring and plug it is difficult to overfuse or defeat the fuse in these circuits.

When a fuse rating larger than 30 A is required, a cartridge fuse is recommended. The ferrule-contact cartridge fuse is avail-

Fig. 1-16. Typical fuses used in home wiring systems.

SCREW-PLUG FUSE

SOLID CARTRIDGE FUSES

FUSE SYMBOL

able in ratings up to 60 A and is divided into two sizes to prevent overfusing. The two-inch ferrule fuse comes in sizes up to 30 A while a three-inch fuse is used for ratings of 31—60 A.

In circuits with current ratings above 60 A a knife-contact fuse is used. These too come in various lengths to ward against overfusing.

Circuit Breakers

Much more convenient and used nearly exclusively in new construction, the circuit breaker is gradually replacing the fuse as a protective device. Circuit breakers perform the same function as fuses and can be obtained with a time-lag feature similar to slow-blow fuses. There are three types of circuit breakers, each employing a different method of operation.

The most common type is the *thermal*. Thermal circuit breakers have a bimetallic element built into the breaker that changes shape as current heats the two metals. Current flowing through the circuit heats these metals. Since they are of different substances they bend at a different rate. The elements act as a latch in the breaker mechanism. When bent they cause the breaker switch to trip.

Magnetic circuit breakers depend on increased current to create a magnetic field strong enough to trip the magnetic switch. The third type is a combination of the thermal and magnetic breakers.

SERVICE TOOLS

Compared to other trades, servicing home laundry equipment does not require many special tools. The majority of jobs require only basic handtools that you probably already own if you are now doing any type of service work. The exact number and kind of tools will vary according to individual preference. A good set of combination wrenches (open-end on one end and box-end on the other), a hammer, pliers, screwdrivers, adjustable wrenches, a socket set, a soldering gun, and a voltmeter are the backbone of this particular box. The voltmeter will be discussed in more detail later. First, let's look at some of the special purpose tools you may need if you intend to service many washers and dryers.

Special Purpose

Like any other professional, the home laundry repairman has need of certain special tools not generally used in other work. These tools represent a considerable investment and should not be purchased without first investigating the need and frequency of their use. Some of these tools are used for only one job. Some, for example, are only needed when a transmission overhaul is called for. If the tool is only a time saver and you do only a limited amount of this type of work, you should consider carefully the cost of such a tool and the actual amount of time it will save you.

Most special purpose tools (Fig. 1-19) are available from the appliance manufacturer. Some service literature will list the tools you should have in your shop and gives part numbers for these devices. Care should be exercised in the purchase of these tools. If you only intend to service one brand of appliance, your choice should be along the lines of tools specified by that manufacturer. When a complete disassembly of a washer is necessary, you will find a spin-basket support (see Fig. 1-19) very useful. These supports vary from brand to brand, however.

Snapring pliers are available in two types and several sizes. One type is for snaprings that must be expanded to be installed or removed. The other type is for snaprings that are compressed for installation and removal. Depending on the make of appliances you service, you may need several sizes of these pliers.

Bearing pullers, gear pullers, and spring compressors also come in many sizes and types. Here again the manufacturer's service literature can offer valuable assistance in making the right choice in purchasing your equipment. This is especially true with brakeband tools. The one shown in Fig. 1-19 is used to expand the brakeband on one make of automatic washer. This tool may not work on some other makes, however.

Fig. 1-18. Tamper resistant fuses are used to prevent over-fusing.

ADAPTER

PLUG

1. SPANNER WRENCH
2. SPANNER WRENCH
3. SEAL REMOVAL AND INSTALLATION TOOL
4. SEAL REMOVAL AND INSTALLATION TOOL
5. BASKET SUPPORT
6. BRAKEBAND EXPANDER
7. DRIVE-BLOCK PULLER
8. RIVET GUN
9. SPRING EXPANDER

Fig. 1-19. Some special purpose tools are needed to repair washers and dryers.

It is possible to make some special tools. In fact, you may even design a few of your own to speed up a job or make a certain service more convenient to perform. One example is a special pair of pliers for removing or installing spring-type hose clamps (Corbin clamps). A groove may be cut into each jaw of a conventional pair of slip-joint pliers to make it easier to hold these clamps in an open position. This tool can also be purchased as shown in Fig. 1-20. It makes it much easier to remove or install spring clamps.

Electrical Tools

Electrical tools associated with washer and dryer repair are of the general purpose type. That is, they can be used for more than one type of work. Some of these tools can also be made in your own shop.

In working with automatic laundry equipment, you will be making two basic electrical measurements—voltage and continuity. The test lamps shown in Fig. 1-21 can be made with little effort or expense and will do a satisfactory job in most cases. A lamp to check for voltage can be made from an old lamp socket and a pair of test leads. A 120 V lamp is connected as shown in the figure. With the appliance connected to the electrical outlet, the test probes can be touched to various points in the circuit to see if voltage is present. If so, the lamp will light.

The other test lamp is used to check continuity in the circuit. Here power must be removed from the appliance and the test lamp is plugged in. The test probes can now be touched to various points to see if current will flow between these points.

A word of caution here. Always use extra care when using test lamps to check a circuit. You are dealing with a live circuit and could easily receive a harmful or even fatal electrical shock.

A good electrical test kit represents a safer and more modern approach to washer and dryer service. For that reason this book will deal only with troubleshooting methods that employ the voltohmmeter (VOM). The meter you select for your toolkit will depend on your own personal preference but it is not necessary to spend a lot of money on an elaborate meter. A meter suitable for appliance repair should have the following features and usually costs less than $50.

- Resistance scales—R × 1; R × 100; R × 10K.
- Ac voltage scales—0—2.5 V; 0—10 V; 0—25 V; 0—100 V; 0—250 V; 0—500 V; 0—1000 V.
- Dc voltage scales—0—0.5 V; 0—5 V; 0—25 V; 0—50 V; 0—100 V; 0—250 V; 0—500 V; 0—1000 V.
- Direct current scales—0—5 mA; 0—50 mA; 0—500 mA.

Some technicians prefer a meter with an ac feature up to 20—25 A. Direct measurements of alternating current are usually much less convenient to make than direct voltage measurements, and very seldom is it necessary to make current measurements. If you find it necessary to make alternating current measurements in your service, a separate ammeter can be purchased for that purpose. This book will show you how to use voltage and resistance measurements to perform almost any type of diagnostic test.

Fig. 1-20. Corbin clamp pliers make it easier to remove and replace spring-type hose clamps.

Fig. 1-21. Test lamps are used to
make voltage and continuity
checks.

To complete the electrical toolkit, we need to add tools for making electrical connections. Most electrical connections in washers and dryers are made with solderless lugs. A good assortment of lugs and a lug crimper should be a part of this kit. In addition, for those few connections that are soldered, a soldering gun of 100—140 W is recommended.

The tools outlined here are the basics—just enough to get you started. As you can see, most of these are contained in a standard toolbox. You will want to add those that you feel make your job easier.

Washer and Dryer Motors

AUTOMATIC WASHERS AND DRYERS RECEIVE THEIR MECHANical power from electric motors. These motors have undergone several changes through the years but have been somewhat standardized in modern laundry equipment. Horsepower varies from one-quarter to one horsepower, depending on the load capacity of the machine. The motor may be single-speed or multi-speed and may or may not be reversible. Almost all laundry appliance motors are now split-phase although some older models still use capacitive-start motors.

These motors usually give little trouble throughout the life of the appliance. Normally, the switches, wiring, and other controls tend to give more problems than the motor itself. Many inexperienced technicians, however, frequently mistake defective motor controls for motor failure.

SPLIT-PHASE MOTORS

Split-phase motors are very simply constructed and consist of only three major parts: the motor frame, the field windings, and the rotor. The frame supports both the field windings and the rotor. Bearings in the endbells of the main frame support the rotor and, at the same time, allow the rotor freedom of motion.

The field windings consist of several coils of wire placed around supporting pole pieces extending from the main frame. Depending on the exact type of motor, there will usually be four or more pole pieces. If the motor is a single-speed type there will be one group of *run* windings, consisting of many turns of heavy wire, and a group of *start* windings, composed of relatively few turns of much lighter wire. Two-speed motors will have two sets of run windings.

A squirrel cage construction rather than windings are used in the construction of the rotor. The rotor consists of a laminated cylinder with slots cut into the surface. The squirrel cage winding usually consists of heavy copper strips pushed into slots cut into the laminated cylinder as shown in Fig 2-1. Each copper strip is brought out to the end of the cylinder and shorted together by a shorting ring.

The electrical relationship between the squirrel cage rotor and the field windings is very similar to that existing between the windings of a transformer. This relationship can be better understood with the aid of Fig. 2-2. If an alternating current is connected to the primary winding of the transformer, a pulsating magnetic field will be established about the primary winding. This field will expand and collapse with each half cycle of the alternating current. As the current reverses, the magnetic field reverses. The pulsating magnetic field induces a separate current in the windings of the secondary.

If a short is placed across the secondary (as in Fig. 2-2) very little resistance will oppose the induced current. Therefore, this induced current will be high. This current will pulsate as the current in the primary pulsates and will likewise create its own mag-

Fig. 2-1. Rotor mechanism of a typical induction motor.

PULSATING
MAGNETIC FIELD

INDUCED CURRENT
(VERY HEAVY
DUE TO
SHORT CIRCUIT)

PRIMARY   SECONDARY

Fig. 2-2. Transformer action in a motor.

netic field. It is the relationship between these magnetic fields that causes the rotor to spin.

Look back at the illustration of the squirrel cage rotor and compare it to the secondary of a transformer. The copper bars in the slots can compare to the windings of transformer secondary. The shorting rings at each end are the shorting jumper at the end of the laminated cylinder. The laminated cylinder serves the same function as the core of a solenoid or transformer: to increase the strength of the magnetic field.

In actual practice the field coils are not wound around individual pole pieces as they are in direct current (dc) or universal motors. They are formed into slots in the frame and arranged to set up a definite magnetic field. It is the interaction of this magnetic field with the magnetic field of the rotor that causes the rotor to spin. The alternating current applied to the field windings goes through its cycle; the magnetic field of the rotor tries to stay up with the moving magnetic field of the field windings. Therefore, rotation is sustained. Note the word *sustained*. Although the run windings can generate the magnetic field necessary to keep the rotor turning, it cannot by itself start the rotation.

The motors used in most automatic washers and dryers usually employ a method of starting rotation that identifies them as split-phase motors. Either capacitance, inductance, resistance, or a combination of these is used to develop the initial starting torque. A split-phase motor has two windings, one *start* winding and one *run* winding, as shown in Fig. 2-3. With the motor at *rest* no power is applied and the start switch is closed. This places the start and run windings in parallel. Since the start winding is made up of several coils of very fine wire, its resistance is much higher than that of the larger wire with fewer coils. This results in a phase difference in the current flowing through the two windings. This phase difference and the fact that these coils are physically located at different points around the frame cause the magnetic fields generated by the current in the coils to set up a strong magnetic field that

START WINDING

120 Vac

RUN
WINDING

MOTOR ROTOR

CENTRIFUGAL
SWITCH

will start to spin the rotor. As soon as the rotor reaches its designed run speed the start switch will open and remove current from the start winding. With the motor already running, the run winding will give the rotor sufficient torque to maintain its speed.

Opening and closing of the start switch is controlled by centrifugal force from the spinning rotor. Action of the centrifugal start switch is illustrated in Fig. 2-4. With the motor stopped or running at a slow speed the weights are held inward by the tension of the springs. In this position the start switch is closed and current can flow in the start winding. As the motor achieves the proper run speed, centrifugal force acts on the weights, causing them to move against the tension springs. This causes the collar to move down the shaft toward the rotor. Springs attached to one contact of the start switch will cause it to open and remove the starter winding current. When power is removed and motor speed begins to fall

Fig. 2-3. A split phase motor has two windings: one start and one run winding.

Fig. 2-4. The actions of a centrifugal motor start switch.

MOVABLE WEIGHT

SPRINGS

SWITCH OPEN

ROTOR

MOVABLE
DISK

MOTOR AT
OPERATING
SPEED

DISK SLIDES IN
WHEN WEIGHTS
ARE MOVED OUTWARD
BY CENTRIFUGAL
FORCE

off, centrifugal force on the weights is reduced and the springs return the weights to their rest position. As the weights move toward the shaft, the collar is forced against the contact of the start switch closing the points. The motor is now in its *rest* position waiting for power to be applied and the starting process to begin. The start switch is located inside the motor on older models. In later models the switch contacts have been moved outside the motor shell, with the centrifugal device still located inside. Mechanical motion is applied to the external contacts by a small lever extending from the centrifugal device through an opening in the motor shell.

The starting method most often used with automatic washer and dryer motors is shown in Fig. 2-5. This is known as a split-phase capacitive-start motor. Operation of this motor is exactly the same as the motor shown in Fig. 2-3, except for the capacitor action in series with the start winding. Adding the capacitor to the start circuit increases the phase difference between the currents in the two windings, thereby increasing the starting torque. On some appliances, but not too often on laundry appliances, you may find a second capacitor in series with the run winding. Figure 2-6 shows the physical appearance of a capacitive-start motor. Notice the capacitor housing on the motor.

In some applications it is desirable to be able to reverse the rotation of appliance motors. To do this with split-phase motors it is only necessary to reverse either the start windings or the run windings, but not both. Figure 2-7 illustrates how this is accomplished in some automatic dryers. With 120 Vac applied between points 1 and 2 the run winding has the necessary current to sustain rotation. When the motor is at *rest*, current can also flow through the centrifugal switch to the capacitor and the reversing switch. If the reversing switch is in the position shown, terminal 4 at the motor start winding is connected to terminal 3 while terminal 5 is connected to the capacitor. This places the capacitor and

Fig. 2-5. A split phase motor with a capacitive start feature.

33

start winding in parallel with the run winding. With the motor connected like this it would start to rotate in one direction.

Again, assume the motor to be at *rest* and the reversing switch shifted to the opposite position from the previous example. The start winding is again in parallel with the run winding. This time, however, terminal 4 is connected through the reversing switch to the capacitor, and terminal 5 is connected to terminal 3. In this position the start winding is exactly reversed from its previous position. The motor will start as it did before but the direction of rotation is reversed.

Fig. 2-6. A typical capacitive-start motor.

WASHER AND DRYER MOTORS

The general features of the various motors we have been discussing are found in most washer and dryer motors. There are, however, a few mechanical and electrical characteristics with which you should be familiar.

Mechanical Features

There are two basic types of construction used in washer and dryer motors; one type is illustrated in Fig. 2-6. The mechanical features of this type allow the motor to be completely disassembled for repairs. The amount of repairs you perform inside the motor shell will depend on your own service operation. Bearing replacement and repairs to the centrifugal device and switch are usually made by appliance service technicians. Major repairs to the motor, such as rewinding, are the job of a service shop.

Four long bolts extending from endbell to endbell through the frame normally hold the motor together. Before removing these bolts, scribe the relative location of the endbells to the main frame with a small chisel. Center the chisel across the joint between the endbell and frame. Tap the chisel with a hammer. Repeat this at the opposite end of the motor. Be very careful to note the exact order in which parts are removed to ensure easy and accurate reassembly.

Once the bolts are removed, the endbells may be removed for access to the internal parts of the motor. Be careful not to damage the centrifugal switch when removing the endbells. If an endbell is particularly hard to remove, it may be tapped with a rawhide or rubber-faced mallet to loosen it. Do not strike the endbells with a steel-faced hammer since they may crack.

Most motors are mounted in the appliance with various types of flexible mounting. One type holds the motor in place by a clip-type mounting spring that straddles a tight-fitting rubber ring over the endbell.

Frequently, the T-frame motor is referred to as a throwaway motor because it is not designed to be repaired by the service technician. Instead, once the technician has determined that the motor is definitely at fault the entire motor is replaced.

T-frame motors are built with the same general mechanical features as the earlier motors except that the endbells are permanently attached to the motor frame with a special bonding material. The centrifugal start mechanism is located inside one of the endbells,

Fig. 2-7. Reversing a capacitive-start motor.

35

and the mechanical motion of the device is transmitted to the centrifugal switch by a small lever extending through a hole in the endbell as mentioned earlier. These motors are not to be disassembled in the field. If mechanical problems or electrical troubles inside the motor housing are encountered, the entire motor should be replaced. In case of electrical troubles always make certain the problem is in the motor and not the switch.

Electrical Features

Three types of motors are used in modern home laundry equipment (classified according to windings and other electrical features): single-speed, two-speed, and reversible. Single-speed motors usually have a four-pole winding arrangement and rotate at about 1700 revolutions per minute (RPM), while two-speed motors have either a four-pole or six-pole winding arrangement and rotate at approximately 1700 RPM and 1100 RPM, respectively.

Usually, T-frame motors include an overload protector inside the motor housing. This device is in series with the windings of the motor and reacts to excessive temperature or current. If either of these conditions exist the protector opens the circuit through the motor windings and stops the motor. Once the overload condition has been removed the protector will close the circuit through the windings after a cooling period of 5 to 10 minutes. Since the overload protector is inside the housing of a T-frame motor, the entire motor must be replaced if the protector becomes defective. Starting and reversing of these motors is controlled by the motor-start switch and other external controls.

Internal electrical features for most laundry appliance motors are shown in Fig. 2-8. Figure 2-8(A) is for single-speed motors. As shown in the schematic, there are two separate windings. One side of each winding is connected to one side of the 120 Vac circuit through the built-in overload protector. Two-speed motors are practically identical except for the low-speed winding as in Fig. 2-8(B). The speed at which this motor runs is determined by the switch selecting either the low- or high-speed windings.

Shown in Fig. 2-8(C) is a three-speed motor. Although not as common as the single- or two-speed motor, this motor will be found on some automatic washers. It is very similar to the two-speed motor except for a third group of pole windings which give the motor an extra-slow speed of approximately 800 RPM.

MOTOR CIRCUITS

To test motors properly the service technician must fully understand how the motor works under normal conditions. This is especially important with throwaway motors because of the replacement

Fig. 2-8. Wiring diagram of three motors: (A) single speed, (B) two speed, (C) three speed.

cost. When determining if the motor must be replaced, first study its operation while installed in the machine. After removing the motor, test it on the bench to make sure you have made a correct judgment. Always doublecheck your test results to avoid unnecessary replacements.

Single-Speed Operation

The single-speed, nonreversing motor circuit is shown in Fig. 2-9. Three leads, labeled C, R, and S in the diagram, extend from the motor and connect it to the external circuit. The common wire (C) is connected to the neutral leg of the 120 Vac circuit. Notice that the common line is connected in series with both the start and run windings so as to break the circuit to both windings under overtemperature or overcurrent conditions. Run winding R is connected directly to the start/stop switch shown in the *stop* position.

A starting capacitor is connected to the start winding, lead S. The opposite side of the start capacitor is connected to a terminal on the motor-start switch. With the motor at *rest*, the start switch connects the capacitor to the start/stop switch as shown. Careful examination of the diagram will show that the start capacitor and start winding are now in parallel with the run winding.

At the moment the start/stop switch is closed, current flows through both the start and run circuits and the series-connected overload protector. With current present in both windings, the mo-

Fig. 2-9. Single-speed motor circuit (nonreversing).

38

Fig. 2-10. Single-speed motor circuit (reversing).

tor will begin to turn. When the motor reaches its approximate operating speed, mechanical motion (shown by the dotted line in the figure) from the centrifugal device will cause the motor-start switch to open. This action removes current from the start winding and allows the run winding to continue to supply sufficient torque to keep the motor rotating at its proper speed.

The major difference between the single-speed nonreversing motor in Fig. 2-9 and the single-speed reversing motor in Fig. 2-10 is the manner in which the run and start windings are connected to the external circuit. Notice (in Fig. 2-10) how the run winding is connected to the external circuit. Current is supplied from the 120 V common through the overload, the run winding, and terminal R1 to the start/stop switch. At point R2, the connection between the overload and the run winding, a lead is run to terminal 1 of the reversing switch. The hot side of the 120 V supply is connected from terminal R1 through the motor-start switch and start capacitor to terminal 2 of the reversing switch.

With the reversing switch in the position shown, terminals 1 and 4 are made and terminals 2 and 6 are made. Therefore, S1 is connected to R2, and S2 is connected to R1. Since the motor is at *rest*, the motor-start switch has connected the start winding in parallel with the run winding. Closing the start/stop switch will cause current to flow through both windings in parallel and the motor will turn. At the designed speed the centrifugal device opens the motor-start switch and the motor continues to rotate in whichever direction it was started.

Assume once again the motor is at *rest* with the start/stop switch open and the motor-start switch closed as shown in the figure. However, this time let's assume the reversing switch is placed in the opposite position, causing terminal 1 of the switch to connect to terminal 3 and terminal 2 to connect to terminal 5. Again the start winding is in parallel with the run winding but S1 is connected (through the capacitor and motor-start switch) to R1; S2 is connected to R2. This is reversed polarity from the preceding example. Under these conditions, if the start/stop switch is closed the motor will start but in the opposite direction. The single-speed nonreversing motor and the single-speed reversing motor operate identically, except for the initial starting action caused by the reversing switch.

Two-Speed Operation

A simplified diagram of a two-speed nonreversing motor is shown in Fig. 2-11. Note the similarities between Fig. 2-9 and Fig. 2-11. The most noticeable difference is the second set of contacts in the motor-start switch. Although it is not a universal practice, many multispeed motors are wired so that, regardless of the operating speed, the motor always starts as if it were to run on high speed. Then, if low speed is selected, when the operating speed is reached, the motor-start switch will cause the current to be shifted to the low-speed winding. This is done in order to take advantage of the higher initial torque generated by the high-speed windings. Since the motor reaches its operating speed in the first few seconds of operation, it is very difficult to notice the shift from high to low speed.

In Fig. 2-11 the neutral line of the 120 V supply is routed through the upper contacts of the motor-start switch and capacitor to the start winding, and through the speed-selector switch and motor-start switch to either the high- or low-speed windings. Once again all of the motor windings are in parallel with each other and in series with the overload and the start/stop switch. Following the wiring diagram in Fig. 2-11, assume the speed-selector switch is in the position shown. Current flows through both the start and high-speed windings. Since the motor-start switch holds the circuit to the low-speed winding open, no current can flow in that circuit.

Fig. 2-11. Two-speed motor circuit (nonreversing).

As proper speed is reached the motor-start switch opens the circuit to the start winding. At the same time the lower set of contacts shifts from the high-speed winding to the low-speed winding. However, since the speed-selector switch is made between contacts 1 and 3, and terminal 2 is open, no current can flow in the low-speed winding. The high-speed winding continues to supply the torque.

Now let's back up to the original conditions with the motor stopped. Shift the speed-selector switch from *high* to *low* speed. In this instance, contacts 1 and 2 are closed and contact 3 is open. When the start/stop switch is closed, the motor will start on the run and high-speed windings just as it did before. However, this time, when the centrifugal start switch changes position, terminal 2 of the speed-selector switch is connected to the low-speed winding. This removes current from the high-speed winding and allows the low-speed winding to take control.

Three-Speed Operation

Three-speed motors are identical in operation to their two-speed counterparts, with the single exception of an additional *run* winding to obtain the third speed. The addition of a third run winding to Fig. 2-11 would make that circuit illustrate the action of a three-speed reversible motor.

MOTOR SERVICING

Motor servicing is an important part of repairing laundry appliances. This servicing includes all the procedures necessary to locate and correct any malfunction in the motor or its associated circuits. Also included is spotting possible troubles and correcting them before major trouble develops. The importance of proper troubleshooting is illustrated by what a service representative of a leading manufacturer of automatic washers and dryers told me. He stated that motors are unnecessarily replaced more often than any other component of their machines. This was verified by checks made on *in warranty* motors returned to the manufacturer for replacement. Although some of these replacements may be made through honest mistakes, far more are made by poor servicing practices. In this topic we have attempted to show how to avoid these unnecessary replacements through good servicing procedures.

All troubles encountered with electric motors used in laundry appliances can be traced to either mechanical or electrical problems. The first step then, after determining that the fault is associated with the motor, is to determine whether or not the trouble is mechanical or electrical.

Mechanical

Mechanical troubles range from minor problems that cause a noise, which may be harmless but annoying, to a frozen mechanical shaft that completely locks the motor. Customer complaints about mechanical problems may include any of the following: noisy operation, motor overheats, motor doesn't run, appliance hums but doesn't run, occasional failure to start, appliance blows fuses, or just about anything else you can imagine. In a later chapter we will show you how to determine if the problem is caused by the motor or some other part of the appliance. Here we will look at methods of locating and correcting troubles once it has been determined that the motor, or its associated components, is the cause of the problem.

To verify that the trouble is in the motor, it should be disconnected from its load. Do this by simply removing the drive belt from the pulley. In some cases the motor shaft extends from both ends of the motor and a belt is attached at each end. Once the load has been removed from the motor, you should check the freedom of the motor shaft by rotating it with your hands. The shaft should rotate freely. Now check shaft endplay by pushing and pulling the shaft back and forth toward the motor. There should be only a slight amount of endplay. Insufficient endplay will cause binding, too much endplay can cause binding, noisy operation, or even failure to start.

Complaints of noisy operation can sometimes be traced to the motor mounting. Look for loose or weak springs in the mounting and hardened or deteriorated rubber mountings that will cause motor vibration or binding. In this instance you would also want to check the condition of the belts used to drive the load and the pulleys. Noise originating in the motor itself will usually be caused by excessive endplay or defective motor bearings. To correct either of these problems, the motor will have to be disassembled. If the motor is the T-frame variety it must be replaced. In motors that can be disassembled, defective bearings can be replaced and excessive endplay sometimes eliminated by installing spacers.

Mechanical checks of the motor itself are limited to these simple operations. If the mounting, pulleys, and belts are in good shape, the rotor is free to rotate, and there is correct endplay, the motor is mechanically in operating condition. We have not eliminated electrical troubles, however.

Electrical

If electrical troubles with the motor are suspected, the motor should be tested while in the circuit. Most electrical problems are traced to the associated circuits rather than the motor itself. For this reason it is imperative that these circuits be thoroughly tested before a motor is replaced. A three-speed reversible motor is used in Fig. 2-12 to represent the manufacturer's service literature. You should first study this literature to familiarize yourself with the operation of the appliance. Determine just what the circuit is supposed to do and verify the operation by running the motor under all possible conditions.

The motor illustrated in Fig. 2-12 is wired for three different rotor speeds in either direction. Therefore, there are six separate modes of operation for this motor. This can be seen by closer examination of the components in the figure. The reversing switch can be set to parallel the start winding with the run windings to cause clockwise or counterclockwise rotation, depending on the polarity of the connection. Notice in the figure that terminal 1 of the reversing switch is connected through the capacitor and contacts 7 and 8 of the motor-start switch to the neutral leg of the 120 V supply. The individual run windings are also tied to the neutral leg through the motor-start switch and the speed-selector switch. Terminal 2 of the reversing switch is connected directly to the opposite end of the run windings at the junction common with the overload protector. Voltage is supplied to these windings through the protector and the start/stop switch.

With the reversing switch in the position shown, contacts 1 and 3 are closed and contacts 2 and 5 are closed. This connects S2 of the start winding to the common connection of the run windings and the overload protector. S1 of the start winding is connected through the capacitor and motor-start switch to terminal 1 of the speed-selector switch. The run windings are connected to the same point through the centrifugal motor-start switch and the contacts of the speed-selector switch.

Placing the reversing switch in the opposite position causes terminal 1 of the reversing switch to contact terminal 4 and terminal 2 to contact terminal 6. Under this condition, S1 is connected to the common connection of the run windings and the overload protector. S2 is now connected through the capacitor and motor-start switch to the opposite ends of the run windings at terminal 1 of the speed-selector switch. This reverses the polarity of the start winding with respect to the run windings and causes the motor to start in the opposite direction.

The neutral side of the 120 V supply is connected to one of the run windings through the speed-selector switch and motor-start switch. When the speed-selector switch is in the *high speed* position, contacts 1 to 4 are connected to supply voltage to terminal 1 of the motor-start switch. In the *low speed* position, contacts 1 to 3 of the speed-selector switch are connected and voltage is supplied to terminal 2 of the motor-start switch. In the *superlow* position, contacts 1 and 2 of the speed-selector switch are made and voltage is supplied to terminal 6 of the motor-start switch.

With *high* speed (terminal 4) selected and the start/stop switch closed, current flows through the high-speed winding and contacts 1 and 3 of the motor-start switch. When the motor reaches the proper operating speed the centrifugal switch opens contacts 2 and 3 and contacts 7 and 8. This removes current from the start winding while current flow remains in the high-speed winding due to the connections between 1 and 3 of the centrifugal switch.

If *low* speed is selected at the speed-selector switch, then contacts 1 and 3 of that switch are closed. Now when the start/stop switch is closed, current flows through these contacts and contacts 2 and 3 of the motor-start switch. As the motor reaches operating speed, the centrifugal switch opens contacts 2 and 3 and opens 2 and 4. Contacts 7 and 8 open at the same time and the motor runs in the direction selected by the reversing switch at slow speed.

Selecting *superlow* speed causes terminals 1 and 2 of the speed-selector switch to make contact. This connects the neutral line to terminal 6 of the motor-start switch. When the motor is stopped the neutral leg is connected from terminal 6 of the motor-start switch through contacts to terminal 9. A jumper from terminal 9

Fig. 2-12. Three-speed motor circuit (reversing).

to terminal 1 connects the high-speed run winding to the neutral leg. As the motor picks up speed the motor-start switch opens contacts 6 and 9 and contacts 7 and 8. This removes current from the start winding and the high-speed winding. This leaves only the superlow-speed winding, connected through 5 and 6 of the motor switch, to the neutral leg.

If you can understand the operation of these circuits, you should have no problem with the majority of those found in automatic laundry equipment. Most are simpler than the one presented here.

We said earlier that troubleshooting the motor and its circuits begins with observing machine operations. Using the circuit just explained, you can see that you must operate the appliance in all six modes of operation. First, select a direction of rotation. Then attempt to operate the appliance in each of the three available speeds. After this, try the opposite direction of rotation and all three speeds. Failure of the appliance to operate properly in any mode will require further investigation.

After observing the operation of the appliance, visually inspect the circuits to locate any electrical trouble. Disconnect the appliance from the power source and remove the necessary covers to inspect the components associated with the motor and its wiring. Look for broken leads, shorted or bare wires, and loose connections. Repair these before going any further. Carefully inspect the mechanical connection between the motor and the centrifugal motor-start switch on motors with external switches. A loose mounting can prevent this switch from operating properly.

During the operational checks make note of which modes seem to be malfunctioning. This will speed the electrical checks made with a voltmeter or ohmmeter. Some technicians still use test lamps to check the circuitry; however, a voltmeter is easier and safer to use. An ohmmeter is safer still, since a mere 3 V potential is used to test the circuits. This book deals exclusively with the safer voltohmmeter (VOM) checks, and they are the only ones recommended here. First let's look at voltage checks.

Since voltage checks are made with power applied to the appliance, care should be exercised to avoid shock to the serviceman or damage to the appliance.

To check the operation of the start/stop switch with power applied to the appliance, set an ac voltmeter to a scale that will read 120 V accurately. Connect one probe to the *neutral* line in Fig. 2-12. Touch the other probe to terminal 2 of the start/stop switch. With the switch open, no reading should be obtained on the meter. When the switch is closed (appliance started) the meter should read about 120 V.

Check the speed-selector switch in a similar fashion. Connect one probe of the meter to terminal 1 of the start/stop switch. Select high-speed operation at the speed-selector switch and touch the other probe to terminal 4. The meter should read about 120 V. Move the switch to the *low speed* position and measure the voltage at terminal 3. Again expect a reading of 120 V. The *superlow* speed position should result in a similar voltage at terminal 2.

Testing the reversing switch can be done with a voltmeter connected between S1 and S2 of the motor-start winding. The reversing action of the switch cannot be checked in this manner if voltage is present when attempts are made to start the motor. Reversing action *can* be checked by observing the rotation of the motor shaft. The voltage at S1 and S2 will only be present during the first few seconds of motor starting. Once the motor is up to speed, contacts 7 and 8 of the motor-start switch are open and the voltage is removed from S1 and S2. Absence of a voltage at these points during the starting phase can mean either the reversing switch or the motor-start switch (contacts 7 and 8) is defective. To isolate such a problem, connect the voltmeter between C on the motor and terminal 8 of the motor-start switch. No voltage here during the first moments of starting points to a defective motor-start switch.

Other contacts of the motor-start switch can be checked by connecting one probe of the voltmeter to terminal C, common to the motor and the run-winding contacts of the motor-start switch. To check the high-speed contacts, connect the second probe to terminal 3 of the motor-start switch and place the speed-selector switch in the *high speed* position. When the motor starts, 120 V should be indicated by the meter. Also, in both low-speed settings voltage should be present during the starting moments of operation but should be removed once proper speed is obtained. In low-speed operation, voltage should be present between terminal 4 of the motor-start switch and terminal C of the motor—between terminals C and 5 for superlow speed.

Voltage checks can be used to check the motor itself. If the motor fails to start, check the voltage across the start windings and the high-speed run winding. With voltage present at both these points, and the motor not overloaded, the motor should start. If not, the motor should be removed and bench tested.

In the circuit shown in Fig. 2-12, if one of the low-speed windings is open the motor will start, achieve speed, and shift to the selected low-speed winding. The speed will drop since the open winding cannot sustain rotation. As the motor slows, the motor-start switch will return to the *start* position and pick up speed once again. This can be detected by intermittent voltage across the selected low-speed winding.

Resistance readings taken with an ohmmeter can be used to duplicate many of the tests performed in the preceding illustrations. Since these tests are made with power removed from the equipment under test, they offer the safest ways of testing electrical components in laundry appliances.

Testing the motor-start switch is very simple. With the appliance unplugged, connect an ohmmeter (set to a low range) between terminals 1 and 2 of the start/stop switch. As the switch is opened and closed the meter should alternately read infinite and zero resistance. If so, the switch is probably working as it should. The speed-selector switch may be tested in a similar manner. Connect one lead of the ohmmeter to terminal 1 of the speed-selector switch and the other to terminal 4 of the same switch. Place the speed-selector switch in the *high* position. The meter should read zero resistance. Move the switch to either of the low-speed positions. An infinite reading should be obtained. Move the probe from terminal 4 to terminal 3. Select the *low speed* setting and observe the meter reading. It should be zero. Move the speed-selector switch to the *superlow* position and the meter reading should change to infinite. Remove the probe from terminal 3 and touch it to terminal 2. Again the reading should be zero, when the switch is moved to any other position, infinite.

To make complete tests of the motor-start switch using an ohmmeter, you must be able to move the switch position from *start* to *run*. This can be done by locating the mechanical connection between the centrifugal device and the switch. Close observation will usually show this connection can be moved using a small screwdriver for a lever. Care should be exercised to prevent damage to the centrifugal device or the switch itself. Connect the ohmmeter between terminals 7 and 8 of the switch. When the motor is at *rest* the meter should read zero resistance. If the switch is manually moved to the *run* position the meter should read infinite resistance. Similar tests can be made between the other contacts of the motor-start switch.

Resistance readings can also be used to check the reversing switch. First check the resistance between terminals 3 and 6 and between terminals 4 and 5. In both cases the meter should read zero. Now place the reversing switch in either position and measure the resistance between terminals 1 and 3. If the meter reads zero, the reading between terminals 2 and 5 should also read zero. Terminals 1 and 4 and terminals 2 and 6 should read infinite. Reversing the position of the motor-start switch should also reverse the readings.

The capacitor can be checked with an ohmmeter set to a scale of $R \times 100\Omega$. Use the shank of a screwdriver with an insulated handle to short the terminals of the capacitor to remove any residual charge. Connect the ohmmeter to the capacitor terminals. The meter should momentarily deflect toward zero and gradually move to a higher reading as the capacitor takes on the charge of the meter batteries. If this happens, the capacitor is probably alright. If the meter reads zero resistance, the capacitor is shorted. If an infinite resistance is obtained, the capacitor is open. Should the meter read any resistance without the needle deflecting toward zero and returning to the reading, the capacitor is leaky and should be replaced.

A word of caution about replacing motor capacitors. Although the capacitors look physically similar to the electrolytic capacitors used in TV and radio circuits, they are electrically different. Always use a motor capacitor of proper value when making a replacement. Never substitute a TV electrolytic capacitor.

Ohmmeter readings are also useful when testing the motor itself. Measuring the resistance of each winding with an ohmmeter can reveal open windings. If all windings read open (infinite), the overload protector is probably open. If the motor has been run or has had power applied during the last few minutes, give the motor

5 to 10 minutes to cool. Then check the resistance readings again. If the circuit is still open, the motor is defective and should be replaced. When making any resistance reading, make sure there is nothing in parallel with the circuit under test that might give false indications on the meter.

Don't overlook the possibility of an open lead in the wiring when troubleshooting for defects in motor circuits. Resistance readings and visual checks can be used to locate broken connections or leads.

The circuit shown in Fig. 2-12 is not meant to represent any particular make or model machine. It is a typical motor circuit similar to many used in automatic home laundry equipment. Using the manufacturer's servicing diagram, the tests used with this circuit can be adapted to almost any motor-driven laundry appliance.

BENCH TESTING AND SERVICING

Once you have determined the motor itself is faulty, it should be removed from the machine for bench testing. There's only a limited amount of bench service that can be done on a T-frame throwaway unit. Aside from cleaning lint from the openings in the frame, replacing rubber mounting pieces, and testing the electrical condition of the motor, little else can be done. Older motors, however, can be opened for service. For that reason we will use an older motor as our example here for bench service.

Operational Tests

The motor should be removed from the machine and tested on the bench for proper operation. During these tests, there will be exposed current-carrying wires and terminals. Be careful not to come into contact with these circuits and prevent any metal from touching them. It is good practice to have a special workbench on which operational tests are made. The top of the bench should be covered with wood, rubber, or insulative tile to prevent accidental electrical shocks.

In order to demonstrate motor testing procedures we will make reference to a specific motor. For this purpose we have chosen a three-speed motor from a popular automatic washer. This particular motor was chosen for several reasons. As can be seen in the photograph in Fig. 2-6, the starting capacitor is attached to the motor frame and wired internally to the motor wiring and centrifugal motor-start switch. Internal wiring connections are shown in the schematic in Fig. 2-13. External voltages applied to the four input terminals control motor speed. The neutral terminal is connected through control switches to the neutral leg of the 120 Vac supply whenever the motor is to be started. Selecting normal operating speed connects the normal terminal to the hot side of the 120 V line. Selection of *low* or *superlow* would connect the proper terminal to the hot side. Operation is the same as that explained in conjunction with Fig. 2-12.

Before making any electrical tests, check the general mechanical condition of the motor once again. Rotate the shaft to make sure the rotor is free and not bound by frozen bearings or a bent cooling fin on the rotor. Check for endplay by pushing and pulling on the shaft. You are now ready for the electrical checks.

Be on the safe side. Start your electrical tests with a complete set of resistance measurements. Set the ohmmeter on the highest resistance scale and measure the resistances between each motor terminal and the motor frame. Any resistance here indicates an electrical leakage between the motor windings and the motor frame. This usually means a breakdown in the winding insulation and, in most cases, will require the services of a motor shop unless you do your own rewinding. The permissible amount of leakage will depend on the particular motor and is sometimes specified in the manufacturer's instructions. Excessive leakage results in poor efficiency, sometimes causing the motor to run hot. On ungrounded appliances this condition can cause electrical shocks.

Measurement of the resistance at the motor terminals is the next step. Exact readings are difficult to predict since the number of turns in the windings will vary with the motor type and manufacturer. These readings will mean more to you as you become familiar with various types of motors and motor-switch arrangements. We must remove the capacitor cover to gain access to the capacitor terminals (Fig. 2-6). Once the cover is removed, one lead should be unsoldered from the capacitor to prevent the parallel resistance of the windings from causing erroneous readings. Touch one meter probe to each of the capacitor terminals. The needle should swing toward zero then slowly climb. Some resistance, indicating capacitor leakage, will be measured but this should be very high.

Connect one probe of the ohmmeter to the normal terminal and alternately touch the other meter probe to each of the capacitor leads. (*Note:* One lead is still disconnected. Make sure you touch this lead and *not* the terminal from which it was removed.) Take note of the two readings. One will be very low, only one or two ohms at the highest, while the other lead may read very high, but not infinite. The near-zero reading indicates you have located the capacitor lead that connects with the motor-start switch. This also means the contacts of the motor-start switch are closed, as they should be since the motor has no power applied. You should have a high reading between terminal B of the capacitor and the *normal* terminal of the motor (Fig. 2-13), since meter current must flow through both the start windings and the normal windings to get from probe to probe.

120 Vac

NEUTRAL

START/STOP

SPEED SELECT

OVERLOAD

START

LOW

NORMAL

SUPER LOW

Fig. 2-13. Three-speed motor nonreversing) circuit.

Leave one probe of the meter connected to lead B and move the other to the *neutral* terminal. You should have continuity—not infinity but a lower reading than you had before, since only the start winding is in the circuit. An infinite reading means an open exists between lead B and the *neutral* terminal—probably an open overload. Zero resistance (seldom encountered) would mean a short across the start winding. If the motor-start switch is operating properly you should have almost zero resistance between any of the three speed-control leads. Use the wiring diagram for the motor with which you are working and check the various windings and switch contacts.

If the motor does not check out at this point there is not much use in actually running the motor. Instead, any defects noted in the resistance or mechanical qualities should be corrected. Once the motor passes these tests, however, it is time to run the motor under its own power with the capacitor reconnected. This can be done with the test box shown schematically in Fig. 2-14. Test boxes like this can be purchased or shop-made. Power is supplied to the test box through a conventional three-prong grounded plug. The ground wire is tied to the case of the test box and to another lead going to the motor. The neutral wire is brought straight through the test box to a second motor lead. A spring-loaded momentary-contact switch and fuse (no more than 30 A) is placed in series with the hot side of the line. Two hot leads are provided to make the test box adaptable to almost any motor wiring scheme.

To use the test box with a one-speed internal motor-start switch, connect the box to the motor as shown in Fig. 2-15. Connect the ground lead to the motor frame (F), the neutral to the motor common (C), and one of the hot leads to the run-winding terminal (R). The second hot lead is to one terminal of the capacitor. Attach a jumper between the second lead of the capacitor and the start-winding terminal (S) on the motor. Either block the motor securely on the bench or have an assistant hold the motor securely in place with insulated gloves during the remaining part of these tests. Plug the power cord into a conventional three-prong grounded outlet. To test the motor it is only necessary to press and hold the test switch.

This same test setup can be used to test the motor capacitor. Should you suspect the capacitor as a trouble source, a substitute can be used for the test.

Interior Checks

When servicing motors that have removable endbells, a certain amount of service and repair work can be profitably undertaken by the appliance technician. Bearing replacements and repairs to the centrifugal motor-start switch are typical. A light chisel and a small hammer can be used to scribe index marks on the endbells to ensure they are properly aligned when the motor is reassembled. After the endbells have been scribed, the four bolts holding the endbells to the main frame can be removed.

Gently tap the endbells with a wood or rawhide mallet until they can be separated from the frame by hand. Once the endbells are removed, the centrifugal mechanism and switch contacts will be exposed. Check to see that the centrifugal switch will operate smoothly by pressing on the trip mechanism. Switch contacts should be clean and free of pits and burn marks.

Fig. 2-14. A wiring diagram of a test box.

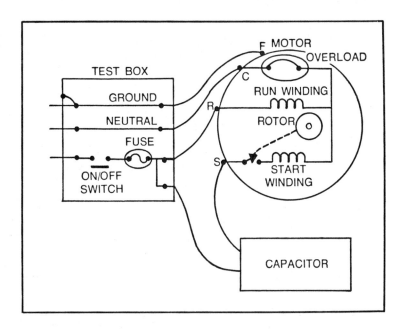

Fig. 2-15. A wiring diagram showing the test hookup for a single speed motor.

Bearings used in these motors are usually of the bushing type and pressed into the endbells. They can usually be removed and replaced by tapping them with a hammer. Exercise care not to damage endbells or bearing surfaces during this operation. In extreme cases it may be necessary to remove these bearings with a bearing press (the type found in electrical motor shops).

When reassembling the motor, take care not to damage the switch contacts or pinch a wire between the endbell and the motor frame.

Automatic Washer Installation

ALTHOUGH MOST OF YOUR WORK WILL BE CONCERNED WITH machines that are already installed, you will find that faulty operation can often be traced to improper installation. Some independent technicians also contract the delivery and installation of laundry appliances for stores not having a service department. This can be a good source of additional income for the small repair shop. Space does not permit installation details for each model of every manufacturer. The general requirements for most machines are given, however, and details can be found in the manufacturer's service literature. From the information given here you should be able to recognize an adequate washer hookup. Make a habit of checking the hookup each time you make a service call. The added sales of hoses, valves, filter screens, and other small parts will pay for the small amount of time it takes to look over the installation.

REQUIREMENTS

The automatic washer must have an adequate supply of both cold and hot water to do the job it was designed to do. Poor water pressure will increase the time required for the washer to complete its cycle. Therefore, proper pressure should be maintained. Water containing sand and other debris can damage not only clothes but the machine as well. A drainage system large enough to handle the water discharge must be provided for removing used water. The water supply must able be protected from freezing.

The weight of the machine itself, combined with the weight of the clothes and water in the machine, requires a solid, level floor for support. If the floor is unsteady, the machine will vibrate and "walk" around. An unlevel floor will result in improper water levels.

Most machines require a 15 A electrical service. If at all possible this should be a separate circuit protected by a 15 A breaker. A three-prong grounded receptacle should be used. In addition to the ground provided by this circuit, it is a good safety practice to have an additional ground between the metal cabinet and the cold-water pipe (if metal pipe is used).

When installing new machines your customer may ask for advice concerning the best location. Convenience to the homeowner should be considered second to operational requirements. Storage area for laundry products and accessories must also be considered. Be sure to talk this over carefully with the homeowner and, when possible, tailor the installation to fit those needs and desires.

LOCATION

Many new homes have specially built laundry rooms with the necessary storage, water, and electrical facilities provided. In this case you will be more concerned that the washer will fit into the specified location. I once knew a very disappointed housewife who discovered that the lid of her new washer would not open without striking the overhead built-in cabinets. Most manufacturers include a chart in the literature accompanying new appliances. These charts show the minimum clearances required for proper installation. If you work with a sales organization, make sure that the sales personnel are aware of these specifications and discuss them with the customer.

Once the location is selected and the washer is delivered, it must be set up. Removing the washer from the shipping crate and disconnecting the shipping braces are usually done at the location, and debris from this uncrating is removed from the home. Packaging and bracing will vary from make to make but usually there are clamps and pads that hold the tub and its suspension mechanism firmly in place during shipment.

The machine requires leveling at the spot where it will sit. Some models have an automatic leveling feature that will maintain a level position without further adjustment once initial setup has been made. Other machines have individually adjusted legs that are varied in height to the desired level. If adjustable legs are provided, it is best not to extend them more than 1 3/4 inches. The longer the legs are extended the less securely the machine will sit.

BASE ASSY.

1 3/4 MAXIMUM RECOMMENDED EXTENSION

BASE ASSY.

LOCKNUT

LOCKNUT

SPRING CLIP (To be Removed)

SPRING CLIP (To be Removed)

FOOT PAD (To be Installed)

FOOT PAD (To be Installed)

REAR ADJUSTABLE FOOT   FRONT ADJUSTABLE FOOT

Fig. 3-1. Adjustable leveling screws on an automatic washer.

Figure 3-1 illustrates how the adjustable legs should be installed. In most cases a bolt is welded to the base of the machine. The legs are screwed into this bolt, and a locknut is tightened to hold the leg in place. In some cases a spring clip is attached to the legs for shipment. This clip should be removed before attempting to level the machine. Rubber foot pads should also be installed under the legs as shown.

A spirit level can be used to determine the correct level. It is best to make two level measurements on the top of the machine. Take one measurement across the width of the machine near the lid and another across the depth. After hookup let the machine fill with water and notice the relative distance between the waterline and the holes in the spin basket. The waterline should be approximately the same distance from all the holes at that height. Try to shake the machine with your hands to make sure it is steady. It is very important that locknuts on all four legs are securely tightened after the washer is leveled and in its final location.

WATER HOOKUP

Automatic washers require an adequate supply of both hot and cold water. The hookup should be equipped with shutoff valves for both supplies. If at all possible the shutoff valves should be accessible after the machine is in place.

Connections between the machine and the faucets are made with rubber hose. Figure 3-2 shows how the hoses should be attached. A rubber washer should be placed in the end of the hose at the machine connection. On the faucet end of each hose a rubber washer and screen should be used. The purpose of the screen is to filter small particles of debris out of the water and prevent them from clogging the inlet valves or other machine components.

Fig. 3-2. Hose connections require washers and screens.

Connecting the drainhose improperly is the most frequent mistake in washer installation. This hose attaches to the machine outlet with a Corbin clamp and connects to the sewer drain. Ideally, the drainhose should fit loosely into the drainpipe as shown in Fig. 3-3A. If the hose fits too tightly, as shown in Fig. 3-3B, it is possible that a syphon will form. The syphon action causes the tub to continually drain and prevents it from filling with water. If it is not possible to allow a loose fit, then an antisyphon valve should be placed in the drain line as shown in Fig. 3-4.

Most manufacturers specify the drainpipe height at somewhere between 32 and 60 inches (Fig. 3-5). Below 32 inches gravity will start a syphon action. Above 60 inches, the discharge pump will be overworked. The flexible drainhose should be looped and permanently fastened as shown in Fig. 3-6.

ELECTRICAL INSTALLATION

Unless you are licensed to make alterations to home wiring systems, you must require that the homeowner place an outlet of sufficient capacity in the immediate area of the machine. Most manufacturers recommend a separate 15 A circuit for the washer. For the best results this circuit should be protected by a 15 A breaker. All washers come equipped with a three-prong grounded

Fig. 3-3. Drainhose connections.

Fig. 3-4. Antisyphon valve pre-vents syphoning at drainhose con-nections.

plug on the power cord. This cord should be connected to a three-prong grounded outlet. The ground wire should never be altered. If a two-prong receptacle is encountered, most manufacturers recommend it be replaced with a three-prong service. When this is not possible (and local codes permit) the electrical hookup may be made with an adapter as illustrated in Fig. 3-7.

Fig. 3-5. Drainpipe height as recommended by most manufacturers.

Fig. 3-6. Drainhose should be looped and fastened securely.

SUDS-RETURN MACHINES

Some machines have a special feature to save hot water and detergent. These machines provide a method of discharging the suds into a tub at the end of the wash period and returning it to the machine to use with a second load. Figure 3-8 shows one manufacturer's recommended hookup for suds-return machines.

Two discharge hoses can be found on these machines. One is a conventional discharge hose, the other is the suds-return hose.

A tub of at least 20 gallon capacity is needed to store the suds during the rinse and spin-dry periods. Suds are discharged at the end of the wash period through the suds-return hose into a holding tub. During the suds-return cycle the sudsy water is drawn back into the machine through the suds-return hose.

After any washer installation, you should run the machine through a complete cycle of operation. Check hoses for leaks and the machine for proper fill, agitation, drain, and spin functions. Also check for vibration. This doublecheck procedure will reduce your callbacks.

LAUNDERING FACTS

Occasionally you will receive calls for washer service when the owner is unhappy about the general operation of the machine. In this instance the machine may go through all the separate operations, however, the customer is unhappy with the results: the clothes do not get clean or too much water is left in the machine. Although these complaints can sometimes be traced to a faulty machine, there are many times, especially if the machine is new, when the problem is incorrect use rather than incorrect operation. Such complaints must be handled with tact so as not to offend the customer. Explaining the machine's requirements and operation will instruct your customers in the proper use of their laundry equipment.

The Laundry Process

The laundry process is a complex operation consisting of both chemical and mechanical actions. Soiling is caused by particles of dirt caught in the fibers of the fabric. Some of these soiling agents will react to soaking, while others will not. Some will soften when heated, others will not. Some will react to chemicals and dissolve when saturated with certain detergents, others will not. The laundering process is a combination of methods designed to remove soil from most fabrics in the average home. Some fabrics will not respond to automatic home laundering, however, and require some other method of cleaning.

In a typical automatic washer the laundering starts by wetting the fabric. Most machines provide a means of adjusting the water temperature to the type of fabric being cleaned. Heavy soil in white or *colorfast* cotton fabrics will require hot water, while delicate fabrics and colors require a cooler temperature. Soils that are easily dissolved in water are loosened as the fabric is saturated with water. Once the clothing has been soaked in water at the proper temperature, chemical and mechanical action is applied to remove other soils. Detergents designed to loosen the bond between the fabric fibers and the soil are mixed with the water. Bleaches, water softeners, and fabric conditioners may be added as needed.

The wash water is now a blend of chemicals and water of correct temperature combined to loosen the soiling agents. Some machines provide a *presoak* period that allows the clothing to sit in this solution for a short time, giving the chemicals an opportunity to work. After the clothing is thoroughly saturated, mechanical action is applied to remove other dirt. Older laundry methods required rubbing or beating the clothing to loosen the dirt. The automatic washer is more gentle and, in most cases, more thorough in this function. A paddle wheel or agitator in the tub oscillates to both move the clothing through the water and circulate the water over the clothing. This causes the clothes to bend and tumble in the water and rubs the fibers in the material against each other. Combined with water circulation in and around the material, this rubbing action loosens dirt caught between the fibers and suspends it in the washer.

The automatic washer allows a specific time for the wash cycle then forcefully expels the wash water and the dirt suspended in it. In most machines this is done by draining the water from the tub then spinning the tub at a high speed to sling the soiled water from the clothing. As this water is drained from the tub, particles of dirt and scum (formed by the chemical reaction of detergents, softeners, and minerals) suspended in the water cling to the external surfaces of the materials being laundered. To remove these the machine is filled once more with clean water. Again they are flexed in the water by mechanical agitation. After rinsing, the water is pumped from the machine and the tub is spun at a high speed to remove most of the water from the clothing.

The clothing should be *damp-dry* when the washing is complete. This is a hard term to define because it means different things to different people. Most machines are built to specifications that leave the clean, damp-dry clothing weighing approximately twice what it weighed before washing. Usually, a fairly accurate check of the machine's water-removing performance can be made by weighing the clothing before and after it is washed.

Water

Automatic washers require a large quantity of water to do their job properly. The exact amount, of course, depends on the amount of washing, the size of the machine, and the types of fabrics washed. An adequate supply of hot water, at 140-170 degrees Fahrenheit, must be maintained for good cleaning.

The hardness of the water will also affect the quality of the wash. Hard water has excessive amounts of mineral deposits that prevent the detergent from working. Also, the detergent reacting with these minerals leaves a scum in the machine that can clog water passages and leave a gray tint in the fabrics. Water hardness is usually expressed in grains per gallon (GPG). Between 0 and 3 GPG, the water is considered soft; 8 GPG is average hard water. Water with a mineral content of 19 GPG or more is excessively hard and requires some sort of treatment to counteract the effects of the minerals.

Chemicals

One of the oldest chemicals used in the laundry process is soap. Although this term is loosely used with all types of cleaning agents, soap is a compound of animal fats and oils. The animal fats combine with the dirt in the fabric and float to the top of the wash water. Soap works well in hot soft water. However, in cool or hard water the minerals and animal fats form a scum that floats on the wash water. When the water is drained from the machine some of the scum clings to the fabrics being laundered. This leaves the fabrics with a dingy appearance.

Detergents are *synthetic* soaps that overcome some of the problems associated with regular soaps. These cleaning agents are chemical compounds of several ingredients with specific purposes. Wetting agents in detergents cause the fabrics to "open up" and allow better saturation. Some of these same agents react with the dirt to keep it suspended in the wash water and prevent it from being redeposited on the clothes. Some detergents contain phosphates to aid in the wetting action and soften the water. Softeners react with the minerals in hard water to suspend them in the wash solution. Stabilizers (boosters) are used to control the sudsing. Various agents used to make fabrics appear brighter are found in many synthetic detergents. Some detergents contain enzymes that react with protein to loosen such stains as chocolate, milk, and blood.

If the water is too hard, much of the detergent is used to counteract the hardness. This reduces the cleaning ability of the detergent and causes a residue similar to soap scum. To overcome this the water can be treated with borax. Borax, as does many softeners, combines with grease and oil in the water to form a soap and softening compound. Some detergents contain borax or other chemicals as softeners. Water-softening chemicals that can be added to the wash water are also available. In extreme cases, however, the only remedy for hard water is a treatment system that will remove some of the minerals.

Bleaches used in home laundry products are either chlorine or oxygen compounds. Bleach will remove soils and stains and whiten fabrics that detergents alone cannot. These chemicals are more harsh than common laundry detergents and should be used with care. They should not be used on wool, silk, or certain blended and noncolorfast fabrics.

In addition to water softeners, there are also fabric softeners on the market. These products make the fabrics softer, fluffier, and easier to handle. They also reduce the tendency of synthetic

64

fabrics to cling due to static electricity.

With today's synthetic fabrics, starch is not used as frequently as it was in the past. Starch is made from a variety of ingredients (cornstarch, wax, and resins, for example) and is used to give a smooth appearance to fabrics. Similarly, *bluing*, once used to brighten the fabrics, is being replaced by fluorescent dye-brighteners in detergents.

LAUNDERING PROBLEMS

Many laundry problems can be overcome by proper installation and proper use of laundry equipment and chemicals. Some common problems and possible solutions follow. Each new machine comes with complete instructions. Always refer to the manufacturer's instructions and those contained on the packages of laundry chemicals. If you suspect that improper use is the cause of a customer's complaint, tactfully ask the customer to demonstrate the problem with an actual wash load. Observe how the clothes are sorted, the machine is loaded, etc. If mistakes are observed use the manufacturer's literature to suggest corrections.

Filters

Many automatic washers contain filters to remove lint. The effectiveness of these filters is reduced if the machine is overloaded, clothing is not properly sorted, the machine is improperly operated (cycle too rough for fabric being washed), the water level is too low, or fabric softeners are used improperly. Fabric softeners should be used only in the final rinse.

Overloading

This will frequently damage the materials being washed. Washers have a definite safe working load that should not be exceeded. Most machines are now designed for 14-, 16-, and 18-pound capacities.

Stains

Some stains when given a chance to set are very difficult, if not impossible, to remove. A stain should be treated as soon as possible to prevent its setting. Most stains can be easily removed by rinsing, thereby diluting the stain while it is still fresh. Any stain-removal chart will give proper instructions if the cause of the stain is known. If the cause if the stain is unknown, the following procedures should help eliminate the stain.

- Because some stains are set by hot water, the spot should

be sponged with cold water or soaked in cold water for at least 20 minutes. If the stain persists, laundry detergent can be worked into it, and then rinsed out. If this doesn't work, a chlorine bleach can be used on white cottons or linens, or hydrogen peroxide on other fabrics. Once again, laundry detergent should be rubbed into the stain and rinsed out with hot water. Wool, silk and non-colorfast fabrics are exceptions to the hot water suggestion. The fabric now can be washed in hot water, using a sufficient amount of detergent.

- If all of the above fails, a grease solvent such as tri-chloroethane can be tried. The area should be sponged thoroughly with the solvent and then rinsed. If necessary, this should be repeated. Launder as usual, if the stain still shows, immersion in a salt water solution for 2 to 3 hours may help. If the garment has metal buttons or zippers, soak only the stained portion. A thorough rinse and rewash should follow this step. Equal parts of liquid bleach and water mixed and applied with an eye dropper to small stains may help. Large stains can be dipped into the solution. The garment should then be rinsed thoroughly and washed in the normal manner.
- If chlorine bleach cannot be used, an oxygen-based bleach should be sprinkled on the stain, then the stain area should be dipped into hot or boiling water. If the stain has been set by age or heat, it may still remain after all of the foregoing treatment.

Grayness

Overall grayness or dull appearance of garments can be caused by redeposited particles of dirt or oily material not emulsified and carried away. This condition is often the result of one or more of the following: Incorrect sorting, insufficient detergent, low water temperature, hard water, or overloading.

Yellowing

Yellowing can be the result of a build up of body oils on the fabric caused by one or more of the following: Low water temperature, insufficient detergent, too short a wash time, long storage of soiled items before laundering, or medications and body oils. This type of yellowing is sometimes very difficult to remove.

Yellowing on certain types of fibers: wool, silk, and certain resin-treated cottons and rayons, can be caused by chlorine-retentive resin. Perhaps too much bleach is being used. If the fabric yellows or retains a strong chlorine odor after being bleached, the fabric can be treated by soaking in a solution of 2 tablespoons of sodium sulfite or hypo-sulfite and 1/2 cup of white vinegar per gallon of water, or a color remover may be used. The fabric should then be thoroughly rinsed.

If the water supply contains a great amount of iron or manganese, these minerals can cause any bleach to turn fabrics yellow instead of whitening them. Almost all detergents will counteract a small amount of these minerals, but in extreme cases a chemical filter in the water supply line will be needed.

Finally, a yellow to brown stain that is caused by a build up of chemical salts from perspiration or detergents may occur. This condition can be treated by putting the affected fabrics through a normal wash cycle with no detergent or water conditioner.

This problem is usually caused by one of the following: Overloading, excessive agitation, improper sorting, water level too low, or improper use of fabric softeners.

Pilling

Pilling, which is the curling of broken fiber ends to form small balls on the surface of the fabric, occurs on polyesters, acrylics, or blends of these fabrics. The pilling is the result of abrasion encountered in normal use and cannot be prevented.

The consumer sometimes assumes the pills to be lint. A careful examination of the fabric will reveal if the problem is pilling or lint. Generally, lint can be picked off the fabric while pills, which are actually a part of the fabric, are very difficult to remove.

It is sometimes possible to remove pills of fabric by brushing the fabric with a stiff bristled brush after each laundering. In extreme cases an electric shaver can be used to remove the pills.

Fabric Damage

This can be of a mechanical or chemical origin.

Clothing damaged by mechanical action will generally be in the form of small arc or angular cuts, pulled threads or round pinpoint holes. Arcs and angular cuts can be caused by clothing being caught between the agitator base and the spin basket. Check the agitator to be sure the bottom is fairly flat. The clearance between the installed agitator bottom and the basket should not exceed 5/16 of an inch. Pulled threads and round pinpoint holes can be caused by sharp projections on the spin basket or the agitator. Check for projections by lightly rubbing a stocking or a similar cloth over the basket and agitator. Be sure to check the upper edge of the basket and the bottom of the tub cover, since clothes often fill with air on the rinse cycle and will billow up and contact these surfaces.

Some tears can occur in normal wash action due to the article being worn thin in areas, or, in the case of curtains and draperies, from exposure to the sun. Holding the article up to a bright light will show worn areas.

Chemical damage is usually the result of improper use of bleach; either it was poured directly over the clothes load without dilution or too great a quantity was used. This type of damage shows up as small holes with frayed edges. The material around the hole lacks body and can be easily torn apart. Bleach-damaged clothing also show up as "aged"—overall, or in spots. They have no tensile strength—and tear very easily. Blacklight indicates areas of bleach damage.

There are other less frequent causes of chemical damage where articles are weakened by medications, ointments, deodorants, and porcelain and toilet bowl cleaners. These items might also leave stains that cannot be removed.

Water Systems

T O AID IN UNDERSTANDING THE OPERATION OF AUTOMATIC washers the entire machine can be divided into three major systems: mechanical, electrical, and water. The water system with its associated components in a modern automatic washer is a complex pumping and control mechanism. Understanding the flow and routing of water within the machine will take you a long way in learning to diagnose troubles in machine operation.

The main purpose of the water system is to fill the machine to the correct level with water of the desired temperature and to remove that water from the machine at the proper time. In addition, depending on the features designed into the machine, the water system may also perform such functions as lint removal, suds storage, and bleach or detergent dispensing. These added features will vary from model to model even within the same brand. Therefore, if you are not familiar with the operating characteristics of a particular machine, consult the manufacturer's service literature. In many cases the owner's operating instructions are also helpful.

COMPONENTS

Many types of valves, pumps, filters, and other water-handling devices can be found in the water system of any automatic washer. Some of these devices are electrically activated while others are mechanically controlled. Defects in these components can result in a variety of troubles. Understanding these functions and operations will help in troubleshooting automatic washers.

Solenoids

Solenoids do most of the work of controlling an automatic washer. These devices are nothing more than magnets that are electrically controlled—electromagnets. Such magnets are made by winding several turns of wire around a core that holds the windings in place. When current flows in a conductor, a magnetic field is established about that conductor as shown in Fig. 4-1(A). Winding this conductor in a coil as shown in Fig. 4-1(B) concentrates the magnetic field in one central location. The strongest part of the magnetic field will be centered in the coil.

This magnetic field can be made to exert a pulling force on magnetic metals such as iron. Figure 4-2 is a simplified diagram of a solenoid. An iron rod inserted in the coil of a solenoid is held down by a spring attached to the frame of the machine in which the solenoid is used. The switch (S1) shown in the *open* position connects the coil to the 120 V power supply. In this condition the spring keeps the iron rod pulled downward. When the switch is closed, the magnetic field generated in the coil exerts a pulling force on the iron rod, striking the spring. If the switch were opened again, the magnetic field would collapse and the spring would pull the rod back to its *rest* position.

The action of the solenoid can be used to control various mechanical components of an automatic washer. An electric clock is used to apply power to the solenoid coil at the proper time. The solenoid can vary in size and pulling strength from a very small coil with only a few inch-ounces of pull to a very large coil exerting several foot-pounds.

Fig. 4-1. Magnetic fields around a conductor when current passes. Notice that in (B) the strongest part of the field is centered in the coil.

Fig. 4-2. Simplified diagram of a
solenoid.

Inlet Valves

One use of the solenoid is operating the inlet valve that supplies water to the machine. Figures 4-3 and 4-4 can aid in understanding the operation of these devices. A flexible diaphragm across
the valve port acts as a seal to prevent water flow when the solenoid is de-energized. The spring forces the plunger downward to
hold the solenoid against the valve seat. Water enters the valve
through the inlet port and is filtered by a small screen that prevents debris from entering the valve. Water, at line pressure, exerts an upward force on the diaphragm that tends to lift the

Fig. 4-3. A water inlet valve in the
open position.

diaphragm from its seat. The spring pressure pushing downward on the plunger holds the diaphragm in place. A small hole in the diaphragm allows water to enter the chamber above the diaphragm. Because water pressure is exerted on both sides of the diaphragm, the amount of spring pressure required to keep the valve closed is reduced. The outlet port of the valve connects to tubing and hoses that route the water to the tub of the machine.

Figure 4-3 shows what happens when the solenoid is energized. The magnetic pull of the coil lifts the plunger off its seat. Water pressure then forces the diaphragm off its seat and water flows from the inlet port. This pushes the diaphragm upward. The water can now flow over the valve seat and out through the flow washer. The flow washer is a special rubber washer with a hole in its center. One side of this hole is rounded (the side facing against the flow of water). This washer is constructed to match the valve to the machine by controlling the rate of waterflow.

The diaphragm lifts from the valve seat and water trapped in the opening above the diaphragm drains through the pilot hole in the center of the diaphragm. This water then flows with the incoming water out through the flow washer and into the inlet tubing and hoses.

Power is removed from the solenoid coil to cut off the flow of water. Spring pressure forces the plunger down, striking the diaphragm and pushing it against the valve seat. Water entering the bleeder hole equalizes water pressure on both sides of the diaphragm, and the spring holds the valve closed.

Typically, these are dual valves having two inlets (hot and cold), two solenoids, and one outlet. Such a valve is shown in Fig. 4-5.

Fig. 4-4. A water inlet valve in the closed position when the solenoid is de-energized.

SPRING

SOLENOID

PLUNGER

VALVE
SEAT

FLOW
WASHER

WATER
IN

FILTER
SCREEN

VALVE
BODY

CLOSED

HOT WATER
SUPPLY

VALVE BODY

COLD WATER
SUPPLY

BOTH CLOSED

Fig. 4-5. The two-coil valve in the closed position. This valve controls the flow of hot, cold or warm water.

Sometimes called a mixing valve, it can control the flow of water to furnish hot, cold, or warm water. Two hoses connect the valve to the hot and cold water faucets located behind the machine. Notice that both inlet valves feed into a common mixing chamber and through a single opening into the machine. This two-coil valve, shown in the *closed* position, has water pressure from each inlet port present at the diaphragms. In this instance no power is applied to either coil, both valves are closed, and no water flows into the machine.

In Fig. 4-6 power has been applied to the cold water coil but not the hot water side. The cold water coil lifts the associated plunger, and water pressure forces the diaphragm from the valve seat. Water can now flow from the cold water faucet through the cold water side of the valve and into the mixing chamber. Hot water flow, shown in Fig. 4-7, is similar to that described for the cold

CLOSED

HOT

COLD

OPEN

COLD OPEN

water side. With power applied only to the hot water coil, the cold water diaphragm remains seated and only hot water enters the machine.

To fill the tub with warm water, both coils are energized and both diaphragms lift. This allows both valves to open and both hot and cold water enter the tub as shown in Fig. 4-8. With both diaphragms off their seats, hot and cold water enter the mixing chamber. Although this water is called warm (hot and cold mixed) the temperature depends on the actual temperature and pressure of the hot and cold water supplies. The warm temperature can be adjusted by the customer at the cold water faucet. Usually, cold water pressure is somewhat higher than hot water pressure. Therefore, best results are normally obtained by opening the hot water valve all the way and turning the cold water valve enough to achieve the desired temperature.

Fig. 4-6. A two-coil valve admitting cold water.

74

The actual appearance of the water inlet valve will vary from make to make and model to model. Usually, they operate similar to the description given here. A variation of this is the three-coil valve shown in Fig. 4-9. In this instance a separate coil is energized for warm water. Notice that both hot and cold water are supplied to the warm-water diaphragm. Small check valves may be built into these ports to prevent the hot and cold supplies from mixing except when warm water is selected. A check valve is simply a device that limits water flow to one direction. Check valve operation is explained later in this chapter.

Some models control water temperature to a finer degree than that provided by the valves just explained. In such cases thermostatically controlled valves are used. These valves are made with a thermal element located in the mixing chamber. When warm water is

Fig. 4-7. A two-coil valve admitting hot water.

BOTH OPEN

selected, water flowing over the thermal element causes it to expand and contract. This regulates the mixing of hot and cold water to achieve the desired temperature. A control permits adjustment of the warm water setting through several stages. Some machines are equipped with a three-coil variation of the thermostatically controlled inlet valve. Regardless, the function is the same.

Service to inlet valves is usually limited to replacement. Although kits that furnish the internal parts needed to repair these valves are available, experience has proven that in most cases it is better to replace the valve rather than repair it. One common trouble developing in inlet valves, especially when the machine is used in an area where the water is very hard, is leaks. Small mineral deposits develop on the valve seats that prevent it from seating properly. This allows water to flow into the machine at all times, even with power disconnected. If water enters when the power cord

Fig. 4-8. A two-coil valve admitting warm water. Both coils are now energized.

is unplugged, the inlet valve should be replaced. These valves may also stick in the closed position.

The valve can be tested with an ohmmeter. Measure the resistance across each coil. Although the exact value of this resistance will depend on the model, it typically runs around 500Ω. A test cord can also be used, but cautiously because hot leads will be exposed.

One method of bench testing is shown in Fig. 4-10. Using a Y-hose equipped with three female fittings, the valve can be connected to a cold water supply as shown in Fig. 4-10. The test leads are two meter probes connected to a conventional extension-cord plug. When this cord is plugged into a wall receptacle 120 V is present, so be careful to avoid touching these probes. With water supplied to the Y-fitting touch the probes to the terminals of one of the coils. That port should open and allow water to flow. This test should be done at a workbench with a sink to catch spilled water. Remove the test probes if water flows. The water should stop. Use the same procedure to check the other coil. This test will prove beyond any doubt whether the valve is functioning properly. However, due to the presence of hot terminals and water, it is somewhat dangerous. A safer method is to leave the valve installed in the machine and use the troubleshooting procedures described later in this book.

The ohmmeter test can be used to check the electrical properties of a valve during bench testing. This is a safe, fast way to locate open coils. A simple test that reveals leaky valves is to blow hard into the outlet port of the valve and note whether there is any leakage at either inlet.

Fig. 4-9. A three-coil valve uses a separate coil for warm water.

Dispensers

Some machines use solenoid valves to control the mixing of detergents, bleaches, or other additives. Typically, these dispensers have a container in which the additive is placed. A solenoid valve between the container and the hose supplying the liquid to the tub prevents the agent from reaching the tub before the correct time. The liquid will gravity-drain into the tub when the coil is energized.

In some cases the solenoid valve routes inlet water to a dispenser. The water then carries the rinsing agent (or other product) into the tub. Regardless, if solenoid valves are used, operation and testing is similar to that of the water inlet valves.

Float and Pressure Switches

Float and pressure-operated switches are used to control the water level in the machine. Although being replaced by the pres-

Fig. 4-10. Testing water inlet valves on the bench.

sure switch on most machines, simple float switches, such as that shown in Fig. 4-11, are sometimes used. The electrical contacts of the switch make and break the circuit to the water-filling components and, in a sense, tell the machine the present water level in the tub.

The float system uses a water column, either the actual water level in the tub or water in an open column as shown in Fig. 4-11, to trip a microswitch. The switch may be normally closed (as shown) or normally open, depending on the manufacturer's design. Normally-closed contacts complete the electrical circuit to the inlet valves when the tub is empty. The water level in the column rises as water enters the tub. When the water reaches the desired level the float causes the switch contacts to open, in this example removing power from the water inlet valves. In other instances where the contacts are normally open, the float causes the contacts to close. This moves the timer from the *fill* position and removes power from the inlet valves. In either case the result is the same— the water is shut off.

A more often used water-level control system is shown in Fig. 4-12. A column of air is used instead of water. Air caught in the tube is compressed as water pushes upward in the tube. Since the air cannot escape, the higher the water level in the tub, the more pressure is exerted on the switch.

Figure 4-12 will aid in understanding how the switch functions. An airtight diaphragm held in place by a spring places a specified amount of pressure on the air in the tube. The water rises causing air pressure to increase, and the force against the diaphragm increases. When the water reaches the proper level, the diaphragm is forced against the movable contact of the switch. In this particular example a two-pole switch is used. This is a common arrange-

Fig. 4-11. A typical float switch used to control the water level in an automatic washer.

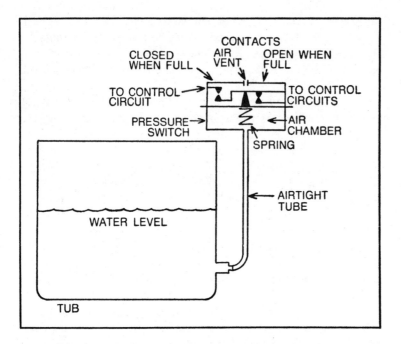

Fig. 4-12. A typical pressure switch is used more often in modern machines.

ment. The hot side of the 120 V supply is connected to the movable contact. When the tub is empty the movable contact is made with the lower stationary contact. At the instant air pressure overcomes the diaphragm-spring pressure, the movable contact breaks with the lower contact and makes with the upper contact. Usually, this action is used to remove power from the water inlet valve and apply power to the timer motor. As water drains from the tub, spring pressure returns the switch to its original position.

There are three principal types of pressure-operated water level controls. The one just described is the single-level type; that is, only one pressure corresponds to a specific water level that trips the switch. A second type has a selector switch with two or more discrete positions that mechanically control spring pressure applied to the diaphragm. This allows the operator to select the desired water level to match the wash load. The third type of control also uses varied spring tension to select water level; however, this switch has an infinite number of selectable levels. The control lever can be set at any water level, rather than only two or three preselected spots, between *low* and *high*.

Don't confuse these operator controls with the calibration control (inaccessible to the user) located on the switch. The calibration control is used at the factory to set the point at which the diaphragm operates the switch. Most manufacturers do not recommend adjusting the level calibration in the field. Instead, if the switch is out of calibration it should be replaced. Repairs or adjustments to these switches are usually short-lived.

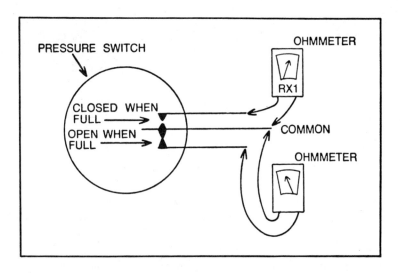

Fig. 4-13. Testing the pressure
switch with an ohmmeter.

Testing these switches can best be done with the switch in-stalled in the machine. After disconnecting the machine from its power source make sure no water is left in the tub. Using the manufacturer's wiring diagram or service data, locate the pressure-switch contacts that should be closed during the filling of the machine. Put the ohmmeter on the R × 1 scale and measure the resistance between the two contacts as shown in Fig. 4-13. It should be zero. Now measure the resistance between the contacts that are open during filling. There should be infinite resistance. Fill the machine to the prescribed level and recheck the resistances. They should be reversed. If not, replace the switch.

Check Valves

Figure 4-14 is a cutaway view of a simple check valve. This valve can be used to allow the flow of liquid in one direction, while preventing it in the opposite direction. The check spring holds the hinged gate against the seat when no pressure is applied. When

Fig. 4-14. A cutaway view of a
typical check valve.

water under pressure is connected to the inlet, the gate raises against the spring pressure and allows the water to flow through the valve. If water pressure is applied to the outlet, the pressure forces the gate against the seat and prevents reverse flow.

Check valves may be tested after removal from the machine by blowing into the outlet port and checking for any leak at the inlet port.

Filters

Water filters used to remove lint from the wash and rinse waters vary from make to make. Some filters, attached to the agitator post, are basket arrangements through which the water is recirculated. The filter basket that catches the lint is a fine mesh screen through which the water passes. The operator removes the basket after the washing is complete and removes the lint.

A more automated version of the recirculating filter is demonstrated with the aid of Fig. 4-15. The check valve near the top of the tub allows water to be drawn from the top of the tub (where the lint is floating) to be recirculated into the bottom of the tub. Lint in the water is trapped by the screen and held in the filter until the water is discharged. During the discharge cycle the pump runs in the opposite direction. Water is now drawn from the bottom of the tub and flows in this reverse direction through the filter. This flushes the trapped lint out of the filter. As the water reaches the check valve the pressure seats the valve in the tub opening and opens the discharge port to route the water out of the machine. In some cases solenoid-operated valves are used to reroute the flow of water.

Pumps

Water enters the machine under water-main pressure. Once the water inlet valve turns off the water, however, the machine needs some method of internally moving the water. A pump is used to recirculate the water in some models during wash and rinse cycles. All machines use a pump to remove the soiled water from the machine. Although these pumps differ in appearance they operate on the same principle.

On some older machines a mechanical device is used to engage the pump pulley with the drive belt during the pumpout phase of the wash cycle. Most modern machines use a pump driven constantly in all modes when the motor is running. During the agitation cycle these pumps are sometimes driven in reverse to prevent pumping. During pumpout and spin cycles the motor reverses and pumping takes place. On other makes a series of check valves and solenoid-control valves determine whether any actual pumping takes place.

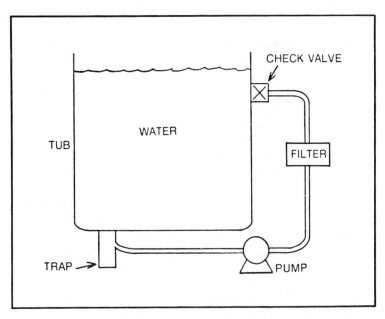

Fig. 4-15. A recirculating water filter is used to trap lint.

Although pump repair kits are available for some models, it is generally a better practice to replace the pump as a unit whenever trouble occurs. The difference in cost between the repair kit and a rebuilt or new pump will usually be less than the labor charge associated with installing a kit. Pump operation can best be checked during actual machine operation.

Some machines are equipped with a so-called two-stage pump. Actually, this is two separate pumps driven by a common rotor. When the pump is turning clockwise, one impeller is turning in the direction to pump water while the other impeller is turning counterclockwise and has no effect on water flow. When the rotation is reversed the opposite impeller begins to move the water while the other turns backward.

TYPICAL WATER SYSTEMS

Although the water systems employed in most automatic washers have many similarities and use many components that perform the same function, there are significant differences from model to model. To use only one or two types for explanation purposes here would shortchange the reader by not providing a broad enough knowledge of water-system operations. Conversely, to cover every type in detail would be beyond the scope of any single volume. What we have attempted here is to give an overall picture of several types of water systems found in today's machines. You should be able to understand the water flow system in any automatic washer by studying either the system in an actual machine or the manufacturer's service literature.

Nonrecirculating

The best system to study first is the nonrecirculating, nonfiltering water system. This is the simplest of all the systems and, since it makes no provision for filtering the wash and rinse waters, it is usually found only on price-leader models. Such a basic system is shown in Fig. 4-16.

Five major components and their associated hoses, clamps, and seals are used in this system. The water inlet valve controls the flow of hot and cold water to the tub. A single hose connects the inlet valve to a small flume at the tub. As selected by the electrical system, the house plumbing admits hot and cold water and routes it through the hose to the flume where it pours into the tub. The tub, sealed at all openings from the mechanical system, holds the water during wash and rinse cycles. The fill control senses the water depth in the tub and, by electrical signal, tells the timing controls when the tub has reached the selected level.

At the time designated by the electrical control system, the pump will begin to remove the water, taking suction at an opening in the bottom of the tub. In order to protect the pump from foreign objects, a sump device is installed at the discharge opening. The sump (manifold, or trap, as it is sometimes called) is constructed

Fig. 4-16. A typical basic water system found in economy model machines.

Fig. 4-17. The sump or trap is used to catch foreign objects before they enter the pump.

Fig. 4-18. A water system timing diagram showing a typical operating cycle.

so that should a heavy object (a small stone, for example) enter the discharge outlet, the weight of the object will cause it to sink to the bottom of the sump. As shown in Fig. 4-17, the pump takes suction above the sump at a distance sufficient to prevent objects from being drawn into the pump. Some of these pumps are built with a removable plug that allows debris to be removed.

A machine using this type of water system would go through an operating cycle similar to that shown in the timing diagram of Fig. 4-18. The complete washing cycle is comprised of three major time periods: wash, rinse, and extract. These periods include one or more discrete operations such as pump, fill, etc. In some cases each period can be divided into smaller time slots that designate specific functions, such as agitate, fill, pumpout, etc. Using the diagram, a typical cycle is explained as follows.

After the clothes are placed in the machine and the detergent is added, the operator starts the machine. At this time the water-

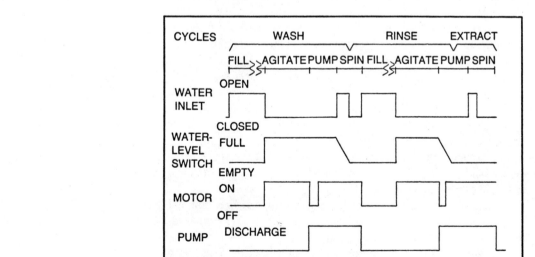

level control senses an empty tub and signals the water inlet valve. Opening the inlet valve starts the fill operation. During this period the only thing happening is water pouring into the tub. The main drive motor is de-energized and the pump is not running.

Upon completion of the filling operation the water-level switch is moved to its full position by pressure in the line from the tub. The water inlet valve is closed and the electrical system signals the mechanical system to start the agitation cycle. The timer is usually started at this point and the machine will wash (agitate) the clothes for a time determined by the timer control.

At the completion of the agitation cycle the electrical controls stop the agitator and start pumping water out of the machine. In this example this is accomplished by reversing the main drive motor rotation which causes the pump to turn in the proper direction and expel water. When the water level decreases to a point where pressure to the water-level control is decreased, the control will shift to the spin cycle. As the spin basket revolves at a high speed, water is slung from the clothes and, since the pump is still operating, pumped from the machine. The length of time that the basket spins is determined by the electrical timer. In most cases the timer will intermittently energize the water inlet valve during this spin cycle. This sprays the spinning clothes with water and removes soap scum and lint. (For simplicity, this intermittent opening and closing of the water inlet valve has been omitted from the timing chart.)

Upon completion of the first spin cycle the driver motor is stopped and the water-level control (now on empty) opens the water inlet valves to fill the machine with rinse water. As during the first fill period, the motor and pump remain inactive. Later we will see that in most machines the timer also does not advance during the fill operation. Most machines have provisions for selecting the rinse water temperature as well as wash water temperature. This is a function of the electrical control circuits and will be explained in a later chapter.

Once the water has reached the correct level, air pressure built up in the water-level control tube switches the control to the *full* position. The timer shifts the solenoids controlling the machines mechanical components to the agitate position. Again the clothes are agitated for a specified period of time. Remaining soil, wash water, and suds residue are removed from the clothes and left suspended in the rinse water. After this cycle the motor is off for a short time.

The timer now starts the motor in the pumpout direction and the pump begins to remove water from the machine. When the water level drops low enough, the water-level control shifts to the empty position, allowing the controls to spin the tub. The pump

86

continues to remove water expelled from the clothes. When the cycle is complete this spinning action continues for a length of time controlled by the timer.

Recirculating

The water system illustrated in Fig. 4-19 has a somewhat different pump, but follows the same general operating cycle as that discussed in conjunction with Fig. 4-16. This particular pump is designed to pump water regardless of which direction the motor turns. The pump is equipped with a valve that may be controlled by a solenoid or the direction of pump rotation. When solenoid controlled, the timer signals the solenoid as to whether the machine is in a wash, rinse, or pumpout operation. The solenoid is energized during wash and rinse operations, and positions the valve to block the discharge line and open the recirculation line.

Water drawn from the tub during wash and rinse cycles is returned to the tub where it exits through a lint filter and re-enters. This filter is simply a strainer that catches the lint suspended in the water. The operator should clean the filter after each wash. In the pumpout periods the solenoid is de-energized. A spring pulls the valve lever in a direction that closes the line to the filter and

Fig. 4-19. A recirculating water system pumps water regardless of the rotation of the motor.

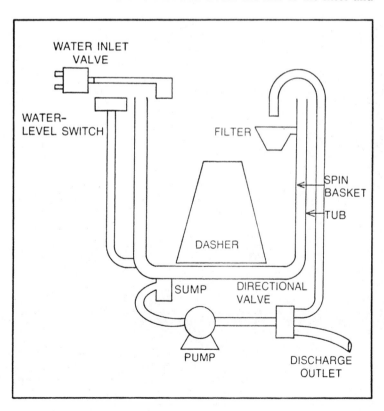

opens the discharge line. Now when the pump rotates, water is expelled from the machine.

Some machines, operating in the same manner, use the direction of the pump of Fig. 4-20. Assume the valve is in position to discharge water from the machine. At the end of the fill operation

Fig. 4-20. Pump-valve operation.

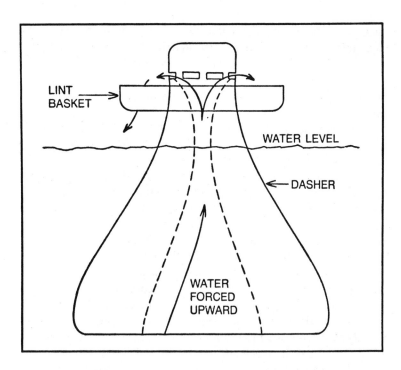

Fig. 4-21. A dasher filter operates
during the wash and rinse cycles.

the motor starts and the pump begins to recirculate the water. The pump starts to rotate and the water, under pressure from the impeller, strikes the valve flipper. This moves the flipper off its set where the impeller can actually engage the flipper and position it to block the discharge hose. The flipper is then held in place by spring tension and water pressure.

The pump reverses rotation to discharge water. Again water pressure moves the flipper off its seat far enough for the impeller to strike it and position it across the recirculation line. Now water is drawn from the tub and returned through the filter.

Dasher Filter

Another filter uses a nonrecirculating water system similar to that in Fig. 4-16. Operation of this system is similar except for the filtering action that takes place during wash and rinse cycles.

When the agitator in Fig. 4-21 is dashed back and forth in the water, the wash (or rinse) water is forced up through the agitator by valves inside the agitator. Water spills out through the upper openings and is strained through the filter encircling the agitator. As with the previous filter, the operator cleans the filter basket after each wash.

Self-Cleaning Filter

Some manufacturers have introduced machines with a self-cleaning filter system. These machines eliminate the necessity of cleaning the lint filter. Figure 4-22 is an illustration of such a system.

This system uses a check valve in the side of the tub just below the normal water level. When pressure is applied to the filter side of the valve, it closes the tub opening and opens the discharge line. Suction applied to the filter side of the line draws water from the tub. Since most lint floats near the surface of the water, these systems are usually more effective than those drawing recirculating water from the bottom of the tub.

The cycle of operation is basically the same as previous systems except for filter operation. When the pump is rotating in the recirculate direction, water is drawn from the tub through the check valve, the filter and the pump, then pushed back into the bottom of the tub. Lint suspended in the water is trapped on the check-valve side of the filter and held until the machine discharges the water. The motor reverses, and the pump is driven in the opposite direction and water is drawn from the bottom of the tub. Water is forced through the filter in the opposite direction. This carries the lint from the filter to the check valve. Water pressure causes the check valve to seal the tub opening, and the lint-laden water is pumped from the machine.

One variation of the system in Fig. 4-22 uses a drive motor turning continuously in one direction. The direction of water flow is

Fig. 4-22. A self-cleaning filter system eliminates the need to clean the filter after each wash.

controlled by a flapper valve in a small manifold attached to the pump. During wash and rinse cycles, the flapper valve routes the water from the check valve through the filter and pump and back into the tub through the trap. This recirculates the water and lint in the filter. During pumpout, the flapper valve reroutes the water in the pump manifold, taking suction at the trap and pumping back through the filter and out through the check valve.

Suds-Return Systems

Some manufacturers offer a special feature on their machines that allows the sudsy water to be stored at the end of the wash cycle and returned to the machine during the next wash load. In addition to special features built into the machine, a suitable tub arrangement must be provided for temporary storage of the sudsy water. Figure 4-23 is a simplified diagram of a complete water system with these benefits.

Initial filling is accomplished using the same methods as other systems. A special two-way valve is inserted in the line between the pump and self-cleaning filter. During the wash cycle, water is recirculated by the pump which draws water through the check valve, filter, and the upper part of the two-way valve, as shown in Fig. 4 -23(A). At the end of this wash phase, solenoid action closes the upper part and opens the lower part of the two-way valve mechanism. When the pump attempts to expel water from the machine in the normal manner, the valve reroutes the water to a hose that sends sudsy water to an external storage tub, as in Fig. 4-23(B).

The electrical timing system repositions the valve for the rinse cycle. Now the system acts like a conventional recirculating filter system as in Fig. 4-23(A). Upon completion of the rinse cycle the two-way valve is held and discharge pumping starts. With the upper port open and the lower port closed on the two-way valve as in Fig. 4-23(C), water is expelled through the filter and check valve in the usual manner.

When the next load of clothes has been placed in the machine and the machine starts to fill, instead of filling through the fill valve in the normal fashion, the pump is started and the lower port of the two-way valve opens as in Fig. 4-23(D). Sudsy water is then drawn from the tub and returned to the machine.

LAUNDRY AGENT DISPENSERS

Several different methods of automatically dispensing laundry additives are used on modern automatic washers. Solenoid valves, mentioned earlier in this chapter, are used by several manufacturers to add detergent or bleach at just the right time. These reservoir-type dispensers consist of tanks emptied (at the proper time)

WATER
INLET VALVE

WATER-LEVEL
CONTROL

DISCHARGE
TO
DRAIN

SUDS
TUB

TUB

SUMP

FILTER

A RECIRCULATION

PUMP 2-WAY
VALVE

2-WAY
VALVE

B SUDS STORAGE

through solenoid valves into the wash water. The tanks are filled by the operator according to manufacturer's instructions. At least two other popular methods are used to achieve the same results. These two systems are discussed here.

Fig. 4-23. The suds-return system allows water to be stored for reuse during the next wash load.

Water Dispenser

Turn to Fig. 4-22 and review the water flow in that recirculating system. Figure 4-24 is a modification of that system that provides a means of automatically injecting additives to the wash water. The major difference between the two systems is the use of a two-

C RINSE DISCHARGE

D SUDS RETURN

stage, single-direction pump with a flapper valve in the upper stage. Actually, the pump is two separate pumps attached to a single shaft. The pump is constantly turning in one direction. The upper stage of the pump functions identically to the flapper-valve pump discussed in conjunction with Fig. 4-22. During wash and rinse cycles the upper stage draws suction at the spring-loaded check valve in the side of the tub, pulls water through the filter, and sends it back to the tub through the trap. At pumpout times the flapper valve on the upper stage causes the pump to take suction at the trap and expel water through the check valve. The check valve seals the opening in the side of the tub and allows water to leave through the discharge hose.

The lower-stage pump delivers a smaller volume of water than the upper-stage pump. During the wash cycle, at the time detergent should be added, the solenoid-operated check valve opens by an electrical signal from the timer. The upper-stage pump moves water from the spring-loaded check valve at the side of the tub and supplies pressure at the trap. The lower pump takes part of this water and routes it through the now-open solenoid check valve to the detergent dispenser. The flow of water through the dispenser carries the detergent to the tub.

Centrifugal-Force Dispensers

One manufacturer makes use of the centrifugal force generated by the spinning basket to inject rinse conditioners into the tub. A donut-shaped dispenser is fitted to the agitator post as shown in Fig. 4-25. The operator loads the reservoir prior to the start of the wash cycle. The rinsing agent is held in the reservoir during the wash cycle. During the wash/spin cycle the centrifugal force of the spinning basket draws the rinse conditioner over the wall

Fig. 4-24. A water circulation detergent dispenser injects additives into the wash water.

Fig. 4-25. The rinse-conditioner dispenser operates on the centrifugal force generated by the spinning basket.

of the reservoir and holds it to the side of the dispenser. As the spinning basket coasts to a stop at the end of the wash/spin cycle, the centrifugal force decreases and the conditioner flows out the bottom of the dispenser.

SERVICING

A faulty water system in an automatic washer is not always readily identified, except when there is leakage. At first you may suspect the electrical or mechanical components as the source of a trouble, but later investigation might prove you wrong. Some technical publications advise you, as your first step in troubleshooting, to isolate the trouble to one of the three major systems. This may be difficult, it not impossible, at times. Suppose the solenoid check valve in Fig. 4-24 is allowing water to be pumped through the detergent dispenser during the spin cycles. Where is the fault? Is it the water system? Is it mechanical? Electrical? The solenoid check valve can be said to be part of all three systems. It is electrically operated (the timer or solenoid coil could be faulty). It is also a device that converts electrical energy to mechanical motion (the linkage could be frozen). And it controls the flow of water (the valve seat could be leaking).

Here we recommend a functional approach to troubleshooting. Treat the entire washer as one system. Break it down into operations such as fill, wash, spin, etc. Observe the cycle of operation and determine what function is not properly completed. Limit your checks to components that affect that function. (This method of troubleshooting is outlined in greater detail in a later chapter of this book—after chapters on the mechanical and electrical systems.)

One of the best troubleshooting aids is a thorough understanding of a machine's operation. When studying any machine pay particular attention to the operating cycle. Knowing what it should do and when it should do it is the key to troubleshooting.

Repairs to water-system components are usually limited, and parts replacement is the most accepted form of repair. Pumps, valves, and water-level switches can be repaired on some models but the cost of labor often justifies replacement with a rebuilt or new component. In our shop, we use a 50 percent rule. Whenever the labor cost runs to 50 percent of the replacement cost, we replace the component rather than repair it. Speed must also be considered. The customer in many cases is more concerned about the length of time required for repair. Availability of replacement parts must then be considered.

Some preventive maintenance checks should be run each time a washer is serviced. Always check the filter screens in the water inlet hoses. Remove any debris that might restrict the water flow. Replace these screens if necessary. Also look for cracked hoses and rusty hose clamps that are a sign of future trouble. Recommend replacement to your customer to avoid callbacks.

Notice the amount of mineral deposits found in the water system when you remove hoses, the agitator, or other parts. This crusty material can cause stuck valves that result in machine overflow. In extreme cases a water softener may be added to the system.

Washer Mechanical Systems

T HE MECHANICAL COMPONENTS OF AN AUTOMATIC WASHER are associated with the drive mechanism. These pulleys, shafts, gears, and clutches move the basket, tub, and agitator. This achieves both the necessary flexing of the fabrics to remove soil and the high-speed spinning to expel water. The automatic washer mechanical system may at first appear to be very complex. If examined closely, however, it is not difficult to understand.

There are two major functions of the mechanical system: to impart agitating motion to the wash and to give spinning motion to the load. To accomplish this a series of gears, pulleys, shafts, solenoids, and belts are used. Together they convert the mechanical rotation of the electric drive motor to whatever motion is needed to accomplish the wash cycle. Two popular systems are now in use to do this job. One is the *tumbler* system, sometimes referred to as a front-loading machine, that uses a horizontally mounted tub. The second is a *top loader* and uses a vertically mounted tub.

TUMBLERS

Tumbler machines are not as popular in home laundries as they once were. There are many of these machines in use commercially however, and some manufacturers still use them for consumer-purchased models. Compared to other methods this system is relatively easy to understand.

Washing action is accomplished by tumbling the clothes in the wash water. This is done by rotating the horizontally mounted tub by means of an electric motor and a pulley and belt arrangement. Water is removed from the clothes by rotating the drum in the same direction but at a much higher speed. The simplified diagram in Fig. 5-1 shows how the tub can be connected to the motor to obtain this operation.

A shaft fixed to the rear of the tub is bearing mounted to the back cabinet of the machine. This shaft supports the rear weight of the tub and its contents. The front of the tub is usually supported by rollers. The motor runs in one direction but has two speeds— one for wash and rinse, another for spin and pump. Mounted on the motor shaft are two pulleys, the main drive pulley and the spin pulley. The connection of these pulleys to the motor shaft is made through a clutch arrangement that mechanically connects and disconnects the pulley and shaft.

The main drive pulley drives a belt that, in turn, drives an idler pulley. A second belt runs from the idler to the main drive pulley on the tub shaft. During wash and rinse operations the clutch on

Fig. 5-1. A tumble-washer drive system.

Fig. 5-2. Pulley clutches are controlled by a solenoid.

the motor pulley is engaged and this pulley drives the tub through the idler system.

A second pulley is used for the spin function. Another clutch-controlled pulley on the motor shaft pulls the belt that drives a second pulley on the tub shaft. The differences in the gear ratio between the two pulley drives and the motor speed account for the difference in the rotation speed of the tub. The clutches are operated by solenoids and determine which drive chain is used to turn the tub.

How these clutches function can be explained with the aid of Fig. 5-2. With the solenoid de-energized, the spring tension exerted on the bellcrank lever keeps the pulley disengaged by sliding it outward on the shaft. When energized by the timer circuit the solenoid pulls upward on the bellcrank. This slides the pulley in on the motor shaft and engages the splines on the pulley with the splines on the shaft. The pulley will then rotate with the shaft.

Although there are differences in the pulley and belt arrangements of tumbler-type automatic washers, most operate in a manner similar to that just explained. Should you encounter differences, they should not give you much trouble if you closely examine the drive system and tub-mounting features.

TOP LOADERS

There are several common features among top-loading machines. All must have some means of converting the rotary motion of the drive motor to the back-and-forth motion needed for agitation. They must also have some method of spinning the basket and its load at a high speed to remove the water.

It is important to understand the three most prevalent mechanical-control systems for top-loading machines. Once you understand these three systems, others should give you little trouble. Exact operation, repair, and service information is given in the manufacturer's literature. If you encounter a machine for which you cannot obtain this literature, study the actual machine and note the operation of each component.

Converting Rotary Motion

Many machines use a drive- and sector-gear arrangement to convert motor rotation to back-and-forth motion. Figure 5-3 illustrates the principles employed in most machines. The mechanism uses four gears and one connecting rod to change the motion at the input of the transmission to the motion necessary at the output. In the illustration these are labeled as the pinion, main drive gear, sector gear, and agitator gear.

The drive motor rotates at a constant speed. Assume that the motor drives the pinion in a clockwise direction (as shown). The pinion meshes with the main drive gear and drives it in a counterclockwise direction at the same constant speed. A connecting rod is bearing mounted at both ends, one end to the sector gear and one end to the drive gear. As the drive gear turns about its shaft, the end of the connecting rod attached to the drive gear is rotated about the drive gear shaft.

The sector gear is also bearing mounted to its shaft. This gear could be compared to a lever with gear teeth at one end and a connecting rod at the other. Its name comes from the fact that it is not a complete gear (360 degrees) but covers only a sector (or portion) of a complete circle. The sector gear meshes with the agitator gear on the agitator shaft. The oscillations of the sector gear will cause the agitator gear to turn first in one direction, then reverse itself and turn in the opposite direction.

In Fig. 5-3(A) the connecting-rod mounting-bearing pin is at the high point of its travel and positions the sector gear at its mid position. As the pinion gear turns the main drive gear counterclockwise, the lower arm of the sector gear is pushed to the left. This moves the sector gear to the right. Since the agitator gear meshes with the sector gear the agitator gear and shaft turns counterclockwise.

In Fig. 5-3(B) the drive gear has made one-fourth of a complete revolution and has moved the sector gear to its extreme right position. As the drive gear continues to turn counterclockwise, the sector gear movement is reversed. When the pin has completed one-half of its revolution as in Fig. 5-3(C) the agitator shaft is at

Fig. 5-3. The sector drive gear provides the necessary back and forth motion.

its original position but moving in the opposite direction. As the connecting rod moves right, the sector gear moves left and the rotation of the agitator shaft is reversed and turns clockwise. This continues until the gear reaches the point where it has completed three-fourths of its revolution as shown in Fig. 5-3(D). At this point the agitator gear would again be reversed and turn clockwise.

Separate Clutch and Transmission

Some automatic washers employ a drive system that includes an enclosed transmission and an external clutch. Most machines of this type house inside the gear case the mechanical components that convert rotary motion to oscillating motion. The transmission and its associated components perform the agitating function. Spin operation is controlled by an external clutch (Fig. 5-4).

The main drive motor is connected to the agitate and spin mechanisms through a series of pulleys, three in this case. Power from the motor is transmitted by a belt to the spin drive pulley and agitate drive pulley. Both pulleys turn whenever power is applied to the motor. The timer will determine which function is to be performed at any given instant by energizing either a spin solenoid, which engages the spin clutch, or an agitate solenoid, which engages the gears within the transmission. Both solenoids are never energized at the same time.

A drive shaft from the transmission enters the tub through a seal. Some tubs have a built-in drive tube through which the drive shaft runs. On others, only a seal at the bottom of the tub prevents water from escaping. When the agitate solenoid is energized by the timer, a gear mesh is made inside the transmission. The drive motor turns, causing the agitate drive pulley to rotate. This drive

Fig. 5-4. An example of an external clutch.

pulley drives a pinion gear as shown in Fig. 5-3. Oscillating motion is applied to the agitator shaft by a sector gear. We now have the familiar back-and-forth motion of the dasher.

During spin periods the agitate solenoid is de-energized and power is applied to the spin solenoid. Spin tension inside the transmission causes the gear mesh at the pinion to be broken. This allows the agitate drive pulley to "freewheel." At the same time, the spin solenoid is energized, bringing the upper and lower clutch surfaces together. Now the power from the drive pulley spins the entire basket. During agitation the spin drive-pulley freewheels about the agitator drive shaft.

This system uses a single-direction motor. The drive pulleys are turned whenever the motor supplies power and always turn in the same direction.

Internal Clutch

Some mechanical systems use a transmission designed around an internal spring clutch to transmit mechanical energy to the dasher and spin basket. A sketch of such a system is shown in Fig. 5-5. The motor transmits mechanical energy to the transmission by means of a belt and pulley drive chain. In this system a bidirectional main drive motor is employed. During the wash/agitate and rinse/agitate cycles the motor turns in one direction (counterclockwise). During spin and pumpout functions the motor direction is reversed. A spring-friction clutch is used to automatically shift the transmission into the mode corresponding to the direction of motor rotation.

Later in this chapter we will examine the workings of the clutch in more detail, but for now assume the clutch operates as stated. As the motor turns in the counterclockwise direction the clutch binds the main drive shaft to the drive pinion as shown in Fig. 5-5. The pinion then drives the main drive gear. A sector gear arrangement, similar to that of the previous system, converts the constant rotation of the main drive gear to the back-and-forth motion necessary for agitation. Seals and bearings permit the agitator shaft to enter into the bottom of the tub and prevent any water leaks around the agitator shaft. Likewise, the agitator shaft and main drive shaft enter the transmission through oil seals.

During spin and pumpout cycles the motor direction is reversed and the motor rotates clockwise. Motion is again imparted to the main drive shaft, but in the opposite direction. The spring-friction clutch now applies enough torque to the pinion to cause the ratchet pawl to engage. This locks the transmission to the case and the entire case is turned by the main drive pulley.

The weight of the transmission and its associated components is supported by a lower bearing in the framework of the cabinet.

Fig. 5-5. An example of an internal clutch.

A solenoid-operated brake is used to hold the spin basket in place during the agitate period. This device is electrically connected to a lid switch that applies the brake to the drive system if the lid is raised during any spin period.

Fluid Drive

Figure 5-6 illustrates the major drive components found in a fluid-type drive. Power from a bi-directional motor is coupled to the main drive shaft by a pulley and belt arrangement. Also fixed to the main drive shaft is the agitator drive pulley which in turn drives the sector drive gear box. Enclosed in the gear box is a pinion and sector gear mechanism similar to that shown in Fig. 5-3.

104

A mechanical clutch inside the gear box disengages the gear box drive pulley during the spin cycle. Spin operation of the basket is achieved by the viscous fluid drive box. Components in the fluid drive box engage the spin drive pulley during the spin cycle and disengage it during agitate operations.

Control signals from the timer determine the direction of motor rotation. During the agitate operation the motor turns in one direction and reverses during the spin operation. When the motor is turning in the agitate direction, the agitate pulley and belt are driving the sector gear box. The internal clutch is engaged and the agitator shaft is driving the agitator back and forth. The spin basket hub is stationary with the agitator shaft moving inside. While the motor runs in the agitate direction the main drive shaft turns freely inside the viscous fluid drive box. The fluid will not allow the spin drive pulley to move.

When the timer reaches the spin cycle, the direction of motor rotation is reversed, causing the main drive shaft direction of rotation to reverse. This reversed rotation is sensed mechanically by the clutch in the sector drive gear box and the agitator shaft is disengaged. At the same time the components in the viscous drive box are attempting to turn against the pressure of the viscous fluid. However, the inertia of the fluid cannot be overcome and the fluid

Fig. 5-6. An example of a viscous fluid drive.

drive box and its attached pulley starts to turn, causing the basket to spin.

TUB CONSTRUCTION AND MOUNTING

The tub components of a typical automatic washer consist of the main water tub, the spin basket, and seals and gaskets at the necessary entrance holes in the tub. Actual mounting of the tub assembly to the cabinet varies. However, the method shown in Fig. 5-7 is similar to that found in many machines. Here the tub is supported on a platform in the lower part of the machine cabinet. The tub is bolted to the platform to rigidly attach it to the framework. Variations of this design omit the platform and use several pieces of channel iron to form a suitable mounting framework.

The components of a typical tub assembly are shown in Fig. 5-8. A main water tub encloses the entire assembly. Most manufacturers use a tub designed with as few penetrations as possible to reduce the chances of leaks. In this example only three penetrations are used. One at the bottom of the tub provides entry for the dasher and spin drive components. A second opening is the trap where water is drawn off. The third fitting, higher on the side of the tub, is a connection for the recirculation hose.

Fig. 5-7. Tub mounting is similar for many machines.

TUB

TUB SUPPORT
PLATFORM

LOWER
MACHINERY
COMPARTMENT

1. MAIN TUB
2. TUB SEAL
3. PERFORATED SPIN BASKET
4. SPIN BASKET HOLD DOWN RING
5. DASHER
6. TUB TO SPIN BASKET RING
7. UPPER SEAL
8. SUMP TRAP
9. RECIRCULATION FITTING

Fig. 5-8. The components of a typical tub assembly.

A perforated basket fits inside the main tub. This basket supports the clothes in the water during agitation and also revolves at a high speed to remove excess water from the clothes during the spin cycle. Seals are used at the bottom of the tub and at the bottom of the spin basket. Tub seals prevent water leakage from the main tub while the gasket and ring attached to the spin basket connect it to the centerpost assembly. Although no water seal is provided for the spin basket, the gasket allows the basket to be tightly connected to the centerpost without damaging the finish.

A bolt at the top of the dasher secures it to the agitator post. The agitator post is driven by the agitator drive shaft inside the spin tub. Separate mounting of the spin basket to the spin tube allows the basket and agitator to be driven by different drive shafts. The large plastic molding at the top of the tub forms a cover between the spin basket and the water tub. This prevents small articles of clothing from falling between the water tub and spin basket.

TRANSMISSION OPERATION

Many physical similarities exist among the components and operations of different automatic washer transmissions. This is not to say that all are identical or that parts are interchangeable. It does mean, however, that the principles discussed here are applicable to most washers.

Solenoid-Controlled

The operation principles of a solenoid-controlled transmission can be illustrated with the exploded view in Fig. 5-9. Back-and-forth agitation is achieved with a sector gear arrangement as previously explained. The main drive gear is turned by the movement of the main drive pulley. A connecting rod moves the sector gear back and forth. No power is applied to the solenoid during the agitation of wash and rinse cycles. Spring pressure keeps the agitator gear meshed with the sector gear. Power applied to the solenoid during spin cycles causes the solenoid to exert an upward force on the agitator fork. This lifts the agitator gear above the sector gear and stops the back-and-forth motion of the agitator shaft.

Fig. 5-9. Exploded view of a solenoid-controlled transmission.

AGITATOR
SHAFT
SPIN
TUBE

TUB

SPIN BASKET
BRAKE ASSEMBLY

MOVEABLE
YOKE

SOLENOID
CAM BAR

ENGAGES
HERE
FOR SPIN

TRANSMISSION

DISENGAGES
FOR
AGITATE

AGITATE GEAR
AND SHAFT

Fig. 5-10. The spinning action is also solenoid controlled.

Spin operation in this washer also uses a solenoid control and is illustrated in Fig. 5-10. This unit mounts on top of the transmission assembly shown in Fig. 5-9 in a manner that extends the agitator shaft upward through the spin mechanism. A spin drive pulley is turned by the drive belt when the main drive motor is running. The spin disc engages the drive pulley only during the spin operation, however. During agitation the solenoid bar holds the yoke up and prevents the spin disc from touching the spin drive pulley. A brake disc on top of the yoke pushes against the spin brake assembly during agitation. This holds the spin tub stationary. During spin cycles the agitate gear is disengaged, as explained in conjunction with Fig. 5-9, and the solenoid in Fig. 5-10 moves the solenoid bar outward. A detent in the solenoid bar allows the yoke spring to pull the yoke down. This releases the brake and, at the same time, engages the spin drive disc with the spin pulley. The spin tube extends up through the main tub and attaches to the spin basket. Spin disc movement is transmitted to the spin tub by the tube.

Self-Contained

The self-contained transmission is designed around the operation of a spring-tension friction clutch. Two views of such a clutch are shown in Fig. 5-11. The assembly is made up of four parts: the main drive shaft, the drive shaft hub (attached to the main drive shaft with a taper pin), the pinion gear and hub assembly (one piece), and the clutch spring.

109

Fig. 5-11. Two views of a spring tension friction clutch.

Only the spring makes mechanical connection between the gear and shaft hubs. The gear is free to rotate about the shaft even though the shaft extends up through the gear. At the point where the shaft extends through the gear, a ring key holds the gear on the shaft. The ring key also holds a pawl mechanism to the shaft. The spring fits tightly over both hubs.

If you view the assembly from the drive shaft end, and the spring is wound in a clockwise direction as shown, turning the shaft counterclockwise would make the spring coils bind together. This would increase the spring tension on the hubs and cause them to act as a solid shaft. Counterclockwise motion of the shaft would then bind the hubs together and the pinion gear would rotate counterclockwise.

When the shaft is turned clockwise, in this example, the force applied to the spring causes the spring to unwind. Since the spring cannot actually unwind, the reverse tension on the coils allows the hubs to slip inside the springs. Torque cannot be exerted on the pinion gear since the hubs tend to slip inside the springs, but the pawl is driven by the extension of the drive shaft through the pinion gear. The pawl is driven only part of a complete revolution until it contacts the main drive gear. The transmission locks when the pawl engages the main drive gear. The clockwise torque on the drive shaft then causes the entire transmission, case and all, to turn clockwise.

This type of transmission is suspended by two bearings. The lower end of the assembly is mounted on a heavy thrust bearing while a similar bearing holds the upper end in place. The drive shaft extends through the lower bearing and attaches to the main drive pulley. The agitator shaft and spin tube extend upward through the upper spin bearing and attach to the agitator and spin basket (Fig. 5-12).

The brake hub is bolted to the bottom of the transmission case. The lower part of this assembly is a machined surface with a brakeband fitted around it. The brakeband fits tight enough to apply friction to the hub but not so tight as to be bound to the hub. In other words, it can slip—slightly—around the hub. As shown in Fig. 5-12 a latch lever is held onto the brakeband by spring tension. This holds the transmission case and spin basket rigid during agitation cycles. During spin cycles the timer energizes the brake solenoid. This lifts the latch from the brakeband and allows the transmis-

Fig. 5-12. Transmission mounting using bearings.

sion case to spin. The electrical circuit for the solenoid usually includes a lid switch in series with the solenoid. Raising the lid will interrupt the circuit and stop the spinning action.

SERVICING

Most manufacturers go to great lengths in designing and constructing transmission systems to ensure a relatively long and trouble-free life. Nearly all automatic washer transmissions are backed by a five-year warranty. Considering the heavy use this component receives, this shows great confidence on the part of the manufacturers. These units are assembled, adjusted, and lubricated at the factory. In many cases they are never opened again for the life of the machine. However, although these are very reliable mechanisms they will occasionally need service.

Lubrication

Automatic washer transmissions are lubricated by a grease bath. The transmission case is sealed and filled with gear grease. Most manufacturers specify SAE 90 weight gear grease. If it becomes necessary to add grease to the transmission, consult the manufacturer's service literature. On many models it will be necessary to completely remove the transmission from the machine.

Adjustments

In early models, transmission adjustments to the gear mesh had to be made internally but few modern transmissions have internal adjustments. Most machines do have mechanical adjustments that must be made occasionally, but they are mostly outside the transmission case. Always refer to the manufacturer's service literature for instructions when making these adjustments.

Brake and clutch mechanisms outside the transmission case will sometimes need adjusting. These adjustments should also be made according to manufacturer's instructions. To ensure proper operation and customer satisfaction, make these adjustments as close to specifications as possible. Take precautions to prevent dirt or oil from contacting clutch or brake linings, since this will interfere with friction coupling existing at such surfaces.

Disassembly and Assembly

These procedures will vary from make to make and model to model. Literature available from the manufacturer will give step-by-step details. Here we will give you a general idea of what is necessary to remove the transmission from a typical machine.

First remove the top cover and console panel. On many

machines they can be removed as a single unit. Quick-disconnect plugs are usually used between the controls on the console and the lower unit components. Spring clips hold the main tub cover in place. To remove this cover it is often necessary to remove the water inlet plume and laundry agent dispensers.

The agitator is usually held in place by a capscrew on the agitator post. After the agitator is removed, the agitator drive block that mates with the agitator must be removed. The spin basket is held to its drive component by a holddown ring and several bolts. Once the holddown ring has been removed the spin basket may be lifted out. Next, the main water tub is removed in a similar manner. Now the transmission can be lifted out through the top of the machine.

On other models, the transmission is removed through the bottom of the machine by first disconnecting the tub and removing the bolts that hold the transmission to the frame. Like transmission removal, transmission-case disassembly will vary from machine to machine.

MECHANICAL PROBLEMS

Mechanical problems in automatic washers may come in one of several ways. Worn bearings, bushings, or gears may show up as noisy operation, sluggishness, or complete binding of the mechanical components affected. Defective seals can be revealed by oil in the wash water or leaks underneath the washer. Troubleshooting the mechanical systems is a simple process in most cases. Usually, these troubles can be located by close observation of the machine's operating cycle.

Some mechanical problems may be corrected in the customer's home. If it is necessary to open the transmission, however, it is recommended that the transmission be taken to the shop. It will be easier to work with the transmission, and there will be less chance of damage to the customer's home. Some larger shops specializing in repair of only a few makes of machines keep a stock of rebuilt transmissions. This lessens the downtime of the machine and increases customer satisfaction. Always check the warranty expiration date when servicing mechanical components. As mentioned previously, mechanical components are usually covered with a long warranty and you may be able to save the customer a considerable amount of money. Most servicemen refer customers with warranty machines to shops authorized to do warranty work on those machines. This may lose you one call, but if you can save the customer some money you'll earn a friend and accomplish a bit of PR for your shop.

During reassembly be sure that the mechanical components are clean and free of dirt and grit. Clean all gears with solvent and dry them with compressed air. When replacing gaskets, use a good-grade gasket sealer and make sure the gaskets are flat and smooth. Seals should be replaced with care. Seal installation tools to reduce the difficulty of seal removal and installation are available from your parts jobber or from the manufacturer.

Electrical Systems

O F THE THREE MAJOR SYSTEMS COMPRISING AN AUTOMATIC washer, the electrical system varies less from machine to machine than any other. Solenoids convert electrical energy to the mechanical motion needed to open and close valves, shift gears, and engage clutches. Timers set the pace for the entire machine operation. These timers are multicontact, multipole switches driven by a small electric motor. Depending on the position of the switch, various solenoids, switches, valves, and even the drive motor is set to the operating condition corresponding to the wash cycle. Solenoids and their operation were explained in an earlier chapter of this book. Here the timer and the overall electrical operation of an automatic washer will be discussed.

MACHINE TIMING

The timer is the "brain" of the automatic washer. Its ability to keep track of time and sequence makes it seem like a very complex component. You will find it can be divided into two separate devices: the timer switch and the timer motor. The switch, as mentioned previously, compares to a multicontact rotary switch. The major difference is the manner in which the contacts are opened and closed. A small electric motor drives an escapement similar to that in a clock. The escapement allows the motor to drive the timer switch in discrete steps, instead of continuously rotating the switch. In other words, the switch moves around in very small increments instead of turning smoothly.

Timers

A simplified diagram of a timer is shown in Fig. 6-1. A series of cams attached to the switch rotor are driven by the timer motor through the escapement. Cam followers are held against the cams by spring tension. As the cam rotates to where the cam follower rides a high spot in the cam, those contacts are closed. When the cam moves to where the cam follower is at a low point, the switch contacts open (see Fig. 6-2). The cams are cut so that contacts and solenoids are energized at the correct time to ensure proper operation of the wash cycle.

Cycles

Timers control machine operation through several cycles. Most machines sold today have at least two different wash cycles. Some have as many as six or eight. No matter how many cycles are featured on the machine, or how many automated operations are ac-

Fig. 6-1. Simplified diagram of a timer.

116

Fig. 6-2. Cam-switch arrangement
of a timer.

complished during each cycle, the time that these operations occur is controlled by a timer similar to the ones discussed in this text.

A close study of the timer dial on any machine will reveal many of its operation features. If you are called to service a machine with which you are not familiar, your first step in servicing should be to examine the timer dial. Note the number of cycles available and the different operations that make up each cycle. Figure 6-3 illustrates some of the features you will see in a typical two-cycle machine.

First, you should notice that there are two separate cycles of operation—a *regular* wash cycle and a *delicate* cycle. The regular wash cycle is for washable fabrics such as colorfast cottons. The delicate cycle is primarily for more delicate materials, such as knits and noncolorfast fabrics. Each of these cycles is divided into smaller increments in which specific operations are carried out. Typically, the regular cycle will encompass 30—40 minutes, while the delicate cycle will be shorter. Similarly, each of the incremental operations is usually shorter in the delicate cycle. This results in a cycle that exposes delicate fabrics to a gentler wash than they would receive in the regular cycle.

Each timer has a definite number of cycles through which it controls machine operations. And each cycle has definite steps. Learning to recognize these steps, and what operations should be completed during each step, is a major part of learning to troubleshoot automatic washers.

Most manufacturers attach a wiring diagram and timing chart to each machine. These diagrams are also included in the manufacturer's service literature. To efficiently repair automatic washers, you not only need copies of these for each machine you service, but you also need to be able to read and understand what the diagrams tell you. If you know how to read the diagrams, you know

117

exactly what the machine should be doing and exactly when it is to do it.

The simplified timing chart in Fig. 6-4 shows the timer sequence employed with the dial shown in Fig. 6-3. Only one cycle is shown in this particular example. The same cycle would be used for delicate fabrics except that each period would be shorter.

Timing operations on most machines start at the end of the fill period as shown in the diagram. Some earlier models have timed periods during which the machine is filled. Most modern machines have done away with this and placed the timer motor in the circuit so that it doesn't start until the machine is full.

Once the water-level control senses that the tub has filled with water, the water inlet valves are closed and the main drive motor is started. The machine continues to agitate the clothes for about 14 minutes during a regular cycle and about 8 minutes in the gentle cycle. Machines with automatic bleach and detergent dispensers release their products into the wash water during the agitation cycle.

At the end of agitation, the main drive motor is stopped and there is a short pause. This gives the drive motor a chance to coast to a stop. Usually, about 1 minute is allowed for this. The timer now engages the proper contacts to start the drive motor for the pumpout operation. Many machines use reversible motors driven

Fig. 6-3. Typical two-cycle timer dial.

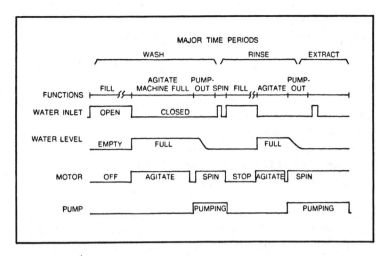

Fig. 6-4. A simplified washer timing diagram whose cycles cor-
respond with the dial in Fig. 6-3.

in one direction for the pump and spin operations and in the opposite direction for agitation. Other machines use a gear mechanism
that changes the gearing arrangements between agitate and
spin/pump times.

The timer would now move from the *stop* position to the
pump/spin position. Mechanical devices shift into the pumpout phase
and water is pumped from the machine. When the water level in
the tub falls to a specified level, the basket starts spinning to remove water from the clothes. Most machines use a spray/spin feature where the water inlet valve is intermittently opened to allow
water into the tub during the spin cycle. The spraying action on
spinning clothes removes soap scum left behind by receding water. Typically, this spin cycle runs about 8 minutes.

A second short pause is encountered at the end of the spin
period. This allows the load to cool for a short time as well as providing a short delay for shifting modes of operation. When energized
by the timer, the machine will begin to fill with rinse water. The
main drive motor is not started until the water-level switch indicates a full tub.

After filling with rinse water, the agitate mechanisms are again
engaged. This works the clothes back and forth through the rinse
water. If the machine has automatic dispensers for rinsing agents,
these devices would be energized during this cycle. This agitate
operation lasts for about 6 or 8 minutes during most regular cycles, and 5 minutes or so during the delicate cycles.

Electrical Operation

Illustrations will be used to explain the electrical operation of
an automatic washer and the timing of various steps. Figure 6-5
is the wiring diagram of a two-cycle washer, while Fig. 6-6 is the
timing chart for the same machine. In actual practice you can

usually find both these diagrams attached to the machine. Understanding how to use them is the key to understanding the machine's operation.

Let's start with Fig.6-5. The common side of the 120 V supply is brought into the electrical system through the push-pull on/off switch. This switch is attached to the shaft of the timer knob. To start the operating cycle the switch must be pulled out. The common side of the line is then brought to terminal 1 of the main drive motor, terminal 1 of the timer motor, terminal 1 of the brake solenoid, and terminal 1 of the water inlet valve. Notice that terminal 1 of the water inlet valve is common in both the hot and cold solenoids.

The hot side of the line is routed to pin 21 of the timer through the overload switch. This is an automatic thermal-overload protector. Whenever current or motor temperature is too high, the overload opens the circuit and removes all power. When the motor current or temperature drops to a safe level, the overload switch will reset and restore power to the circuit.

Now refer to the timing chart in Fig. 6-6. Note how the chart is laid out. This is typical of many machines. Down the side the cams that control the various functions are listed. Eleven cams, A through K, are used. Cam A is used to control water temperature. This switch signals the water-temperature switch as to whether the machine is in a wash or rinse period. Cam B designates agitate or spin functions in a similar manner. As you continue to read down the chart, the function controlled by each cam is listed.

Both regular and gentle cycles are shown in the diagram. The top of each line shows the operations for the regular cycle, while the lower portion shows gentle cycle functions. The specific periods of operation are listed to the right of the cycles. Water-temperature and speed-selector switches are shown below the wiring diagram. The speed selector cam K is a separate cam positioned to close terminals S1 and S3 during the normal wash cycle and terminals S1 and S2 during the delicate cycle.

The water-temperature switch is a separate control positioned by the machine operator. Three temperature selections are provided: hot water wash/warm water rinse; warm water wash/cold water rinse; cold water wash/cold water rinse. On some machines this feature is built into the timer, and water temperature is selected on the basis of the cycle. In this example we use a separate switch.

To operate this machine in the normal cycle, the operator selects the desired temperature, loads the machine with clothes and detergent, puts the timer to the start of the regular or gentle cycles and pulls out the knob. The timing diagram shows cam A is now in position to select wash water. Each of the other cam positions are shown below.

Fig. 6-5. Wiring diagram of a two-cycle washer.

For the remainder of this discussion it will be necessary to refer to both Fig. 6-5 and Fig. 6-6.

As soon as the operator starts the machine, the common side of the 120 V supply is connected as previously explained. At the same time, the hot side of the line is connected through the overload switch to pin 21 of cam J. This is sometimes called the on/off cam since it provides a means for the timer to turn off the machine. In this *off* position, the switch is in the position shown. In the normal cycle pin 21 is made with the upper contact, while the delicate cycle closes pin 21 with the lower contact. Regardless of which cycle is selected, cam J applies power to pin 18 of the timer. At the end of the cycle, cam J removes that power.

Power is routed from pin 18 through parallel circuits to the water-level switch. One of these paths is through the lid switch, the other through cam H. Pin 18 of cam J is common with pin 18 of cam H. At this time the machine is in the wash/fill cycle. Therefore, pins 18 and 8 of cam H are closed. Power is applied to terminal 1 of the water-level switch, even with the lid open. Since the tub is empty, terminals 1 and 2 of the water-level control are closed, applying power to pin 11 common to cams E, G, and I.

According to the timer chart (Fig. 6-6), cam E holds its associated contacts open during the wash/fill function. Therefore, these contacts are not significant at this time. Cam G, as shown by the timing chart, is also inactive at this time. Cam I is active, however, and closes pins 11 and 20 applying power to terminal 1 of the water-temperature switch.

Assuming that cold water has been selected for the wash operations, terminals 1 and 4 of the temperature switch are closed. This routes power back to the timer at pin 3 of cam A. During the wash/fill function, cam A closes pins 1 and 3 supplying power to the cold water solenoid. Should hot water be selected, terminals 1, 2, and 3 of the water-temperature switch are closed. Power is applied to the hot water solenoid through terminals 1 and 2 of this switch. Terminal 3 of the water-temperature switch would supply power to pin 2 of cam A. This cam holds pins 1 and 3 closed and pin 2 open at this time, however.

When warm wash water is selected, terminals 1, 2, and 4 of the water-temperature switch are made. Terminals 1 and 2 energize the hot water solenoid. Power from terminal 4 routes back through pins 3 to 1 of cam A to turn on the cold water.

Although the common side of the power source is connected to the timer motor at this time, the hot side is held open. According to the timer chart, cam C is closed from pins 9 to 7, and cam F is closed from pins 13 to 7. This means that before the timer motor can start, or power can be applied to the main drive motor, terminal 3 of the water-level control must be hot.

Fig. 6-6. Typical timing diagram.

CAM	FUNCTION	PINS	WASH				RINSE				COAST	OFF
			FILL	AGITATE	PAUSE	SPIN/SPRAY	FILL	AGITATE	PAUSE	SPIN/SPRAY		
A	WASH SELECT	3										
	RINSE SELECT	1										
		2										
B	AGITATE	4										
	SPIN	6										
		5										
C	TIMER AGITATE	7										
	TIMER SPIN	9										
		8										
D	SPIN	4										
	AGITATE	10										
		5										
E	BRAKE	11										
		12										
F	MOTOR AGITATE	13										
	TIMER MOTOR BYPASS	7										
		14										
G	RINSE FILL	11										
	RINSE SPRAY	15										
		16										
H	MOTOR SPIN	17										
	DOOR SWITCH BYPASS	8										
		18										
I	WASH FILL	16										
		11										
		20										
J	ON/OFF	18										
		21										
K	CYCLE SELECT	S3					NORMAL CYCLE					OFF
		S1					GENTLE CYCLE					
		S2										

123

The brake solenoid controls a brake mechanism used to prevent the spin basket from turning during agitation. During spin cycles the solenoid is energized. This lifts the brake and allows the spin basket to turn. As we have already pointed out, pins 11 and 12 of cam F are now open and the brake is applied.

As you can see, the machine will remain in this condition until the water-level switch indicates that the water has reached the proper level. Many machines have a variable control to select the water level. This control functions in the same manner as the one shown here. Mechanical features of this switch vary the spring tension, allowing for water-level selection. This added feature does not affect the electrical circuits and, from a wiring diagram, you cannot distinguish between fixed and adjustable water-level controls.

The water-level switch closes terminals 1 and 3 and opens terminals 1 and 2 when the machine fills. This applies power to pin 7 of cams F and C. Cam C has pins 7 to 9 closed, starting the timer motor. At cam F, pins 13 and 7 are closed and power is applied to terminal S1 of speed-selector cam K. Speed is controlled by the cycle selected. During the normal wash, terminals S1 and S3 of the speed-selector cam are closed, placing power on the high-speed winding of the motor. For delicate cycles terminals S1 and S2 of the speed-selector switch are closed and power is applied to terminal 5 of the motor. Notice that terminal S3 of the speed-selector switch is connected to pin 4 of cams B and D. These are the spin and agitate cams. During agitation cycles, cam B closes pins 4 and 6 while cam D closes pins 5 and 10. To control spin and agitation, these two cams are used to determine the direction of rotation for the main drive motor. From an earlier chapter you learned that to reverse the direction of rotation of an induction motor it is necessary to reverse the polarity of the motor-start winding. That is accomplished with cams B and D. Terminal 4 of the motor-start winding is connected to pin 10 of cam D, while terminal 2 of the motor-start winding is connected to pin 6 of cam B. During agitation cycles, the start winding is connected from terminal 2 of the main drive motor to pin 6 of cam B, to pin 5, and to the common side of the line. Terminal 4 is connected to pin 10 of cam D, to pin 4 of the cam, and to the hot side at terminal S3 of the speed-selector cam.

Notice that the connections to terminals 2 and 4 of the start winding will be reversed when the cams shift to a spin function. Also note the internal centrifugal switch of the motor. When the motor is at *rest*, the switch contacts are made as shown. When the motor reaches operating speed, the switch opens terminal 4 of the start winding and terminal 3 and 5 of the run circuit. At the same time, pins 5 and 6 at cam B are closed. This means the motor is always started in the high-speed (regular cycle) mode. If low speed (delicate cycle) is selected, terminals 1 and 2 of the speed-selector switch are closed, applying power from pin 13 of cam F to terminal 5 of the motor. When the motor reaches operating speed, the centrifugal switch will remove power from the high-speed winding and apply power to the low-speed circuit.

This particular machine has a 10-minute wash time during the regular cycle. Notice on the timing chart that all mechanisms remain in their established conditions until the wash period terminates at the 10-minute mark. At that time, cams D, F, and I change their contact positions. Cam D opens pins 5 to 10 and closes pins 10 to 4. This shifts terminal 4 of the motor-start winding from the common side of the supply to the hot side. Cam F opens pins 13 to 7, removing power from the speed-selector switch and the main drive motor. Cam I opens pins 11 to 20, removing power from terminal 1 of the water-temperature switch. A short pause is allowed for these actions to take place. After about one minute, cam B opens pins 4 to 6 and, at the same time, closes pins 6 to 5. This action connects terminal 2 of the motor-start winding to the hot side of the line at terminal 3 of the speed-selector switch. Reversing the polarity of the start winding is now complete.

At the end of the pause, cam E closes the pins to the brake solenoid, unlocking the brake from the spin basket. This starts the spin/spray operation, which lasts for approximately six minutes. Reading across the timing chart, you can see that cam A has both sets of pins open. Cams B, C, and D are in the spin position and cam E has removed the brake. Cam F, during the first few moments of the period, closes pins 7 and 14. This ensures that power continues to the timer motor when cam C is changing the position of its pins. Notice that this happens each time cam C changes position. Cam G closes pins 15 and 16 for a short period. This ties terminal 2 of the cold water solenoid to the hot side of the line at terminal 2 of the water-level switch. When the water is down to a level that causes the water-level control to switch from the *full* position to the *empty* position, the cold water solenoid is activated and a spray of cold water enters the tub.

Upon completion of the spin/spray cycle, cams B, C, and D shift back to the agitate function and the fill circuit is energized to fill the machine with rinse water. Cam A now connects pins 1 and 2, and the water-temperature switch will control the water inlet valve by energizing either the cold water solenoid or both hot and cold water solenoids, depending on whether warm or cold water has been selected for rinsing.

Most machines have a built-in safety feature that prevents the spin operation from taking place when the lid is open. Looking at Fig. 6-5, you will notice the lid switch in the circuit between pin

18 of cam J and the water-level switch. Should the lid be opened during the spin function, the lid switch will remove power from terminal 1 of the water-level control. This removes power from the main drive motor and the brake solenoid. The main drive motor stops and the brake is released, bringing the machine to a sudden stop.

The remainder of this cycle is similar to the operations already described. The cams continue to shift the switches back and forth to control the various solenoids. Delicate cycles are completed almost identically to regular cycles except for the time allowed for each function and the speed of the main drive motor. Always study the timing charts and wiring diagrams of new machines you service. Notice changes or variations with which you are not familiar. Trace out the circuit and refer to the timing chart to determine how these operations are accomplished.

TESTING ELECTRICAL COMPONENTS

Voltage and resistance measurements can be used to check most electrical components of an automatic washer. Test lamps can be made or purchased to test these devices. We recommend the use of a VOM, however. Test lamps require "hot" leads that can damage the machine and expose the technician to electrical shock. In all instances in this book we will describe tests that take voltage and resistance measurements using a VOM. When possible, resistance measurements are preferred over voltage checks. Resistance checks require no outside power source, thus removing the danger of shock.

Wiring Checks

The wiring hookups inside the washer will give the technician very few problems, mainly because failures seldom occur in this area. Usually, failures here are due to loose connections, chaffed or broken wiring, or improper wiring caused by inexperienced personnel attempting to make repairs. Symptoms of wiring failures are too varied to list, and range anywhere from opens and shorts to shocks.

Many failures encountered in the wiring can be located by simply going through the wiring harness and looking for chaffed spots, and loose or broken leads. If you suspect the wiring, use a diagram and check the wiring from point to point. Check continuity from point to point. These tests will usually reveal any fault in the wiring. Voltage tests can be made; however, power must be applied to the machine and this creates a shock hazard.

Solenoids

Solenoids can be tested using a combination of voltage and resistance measurements. As an example, assume that a machine fails to fill with cold water. A voltmeter can be connected across the cold water solenoid as shown in Fig. 6-7. Select cold water wash and rinse with the water-temperature switch and observe the meter for voltage at the solenoid. Repeat the test for each fill position of the timer and each position of the water-temperature switch that activates the cold water solenoid (warm and cold in both wash and rinse cycles). If voltage is present during any of these cycles, and cold water still does not enter the tub, then the solenoid is likely to be at fault. Should you not be able to find any voltage at the solenoid, look for trouble in the timer or water-temperature control.

Resistance measurements are also effective for checking the electrical properties of the solenoid. These checks will not reveal difficulties such as sticky mechanical parts, however. To make resistance checks, remove all power from the machine and set up an ohmmeter for $R \times 10$ operation. Connect the meter across the solenoid. Make sure there are no parallel circuits to interfere with the meter reading. The exact resistance of the solenoid may vary, but in most cases it runs several hundred ohms. An open coil will be indicated by an infinite meter reading, a shorted coil by a very low reading.

Water-Temperature Switches

As with other components, voltage tests of water-temperature switches must be made with the switch installed in the machine and power applied. To test the water-temperature switch in Fig. 6-5, connect one end of the voltmeter to the common terminal of the water inlet valve (terminal 1). Set the timer to the *wash/fill* position, and set the water-temperature control to *cold wash*. Touch

Fig. 6-7. Voltage tests on a solenoid valve.

the other probe of the voltmeter to terminal 1 of the water-temperature switch. Voltage should be present here, indicating that the timer is applying voltage at the proper time. Move the probe to terminal 4 of the switch. Voltage here indicates the switch is capable of selecting cold water. Change the switch to *warm wash* and check for voltage at terminals 2 and 4 of the switch. If voltage is present here the warm water contacts are working. Hot water selection can be checked by moving the water-temperature switch to *hot* and checking for voltage at terminals 2 and 3.

Ohmmeter tests can be made by removing all power from the machine and connecting one probe to terminal 1 of the water-temperature switch. When set for cold water wash, the resistance between terminals 1 and 4 should be zero, and between terminals 1 and 2 and terminals 1 and 3, it should be infinite. Selecting *hot water* should give zero resistance between terminals 1, 2, and 3 and infinite resistance between 4 and any of the others. When *warm wash* is selected, zero resistance should be measured between terminals 1, 2, and 4 and infinite resistance between these three and terminal 3. Should the switch fail any of these tests, replacement is necessary.

Water-Level Control

Voltage measurements to check the water-level control are made by first setting the timer to the *wash/fill* position. Attach one probe of the voltmeter to the common connection of the water inlet valve (terminal 1). With power applied to the machine, touch the other probe to terminal 1 of the water-level control. Lack of voltage here indicates trouble elsewhere in the electrical system. With voltage present at terminal 1 of the water-level control, and the machine in the *wash/fill* position, move the probe from terminal 1 to terminal 2. Voltage should be present at this terminal. Voltage should appear at terminal 3 when the machine is filled with water, at which time there should be no voltage at terminal 2.

Resistance measurements must be made with power removed. Attach one probe of an R × 1 ohmmeter to terminal 1 of the water-level control. If the tub is only partially filled, you must determine if the machine stopped during the fill or drain function. If the machine stopped during the fill operation, there should be zero resistance between terminals 1 and 2 and infinite resistance between 1 and 3. If it stopped during the pumpout phase it will be difficult to determine exactly which way the switch is made, to *fill* position (1 and 3 closed) or the *empty* position (1 and 2 closed). In this case, drain the water from the machine and start the test when the machine enters the wash/fill mode. Here terminals 1 and 2 are made (zero resistance) when the tub is filled with water. (This can be done with a hose to prevent applying power to the machine.) Terminals 1 and 2 should be open and terminals 1 and 3 should be closed.

Lid Switch

Most machines on the market will employ one of two types of lid switches. One popular type is a mercury switch where a small drop of mercury makes electrical contact with two wires inside a small glass ball. When the lid is raised, the lid switch is moved to a position that prevents the mercury from touching one or both wires. This opens the electrical circuit.

Another method is to trigger a small microswitch with a plunger on the lid. When the lid is closed, the plunger strikes the switch causing it to close. When the lid is open, spring tension opens the microswitch.

To test these switches, attach one probe of a voltmeter to terminal 1 of the water inlet valve (or any other point on the common side of the line). Attach the other probe to terminal 1 of the water-level control. With the lid closed, set the timer to the *wash/fill* position and apply power. You should read 120 V on the meter. When the lid is opened, the meter should read zero.

To make resistance checks, remove all power and clip one meter lead to terminal 1 of the switch and the other to terminal 2. Raising the lid should result in infinite resistance. Closing the lid should cause the meter to show zero resistance.

Timers

Testing the timer appears to be, for the inexperienced technician, the most difficult of all tests. However, if you are careful and have a copy of the timing chart and wiring diagram, the tests are not so difficult. Most importantly, keep in mind that you should have some valid reason for testing the timer. For example, suppose a washer fails to fill during a particular cycle. This could be the result of a failure in any one of several areas, including the timer. Eliminate the other possibilities, such as a defective solenoid valve, pinched inlet hose, or defective water level control. After eliminating these possibilities, the timer should be tested.

Voltage tests of a timer are sometimes difficult because the cramped space in the machine console makes it very easy to short out some of the circuits. Always exercise caution when making voltage tests.

Figure 6-8 illustrates the motor-start circuit of the washer shown in Fig. 6-5. Assume it has been determined that the motor is not starting for the spin function while the agitate function seems to be operating properly. To determine if the timer is the cause of the problem, examine the timing diagram of the machine (Fig. 6-6) closely.

Fig. 6-8. Motor reversing circuit.

Remember: in this machine the motor runs in different directions for the agitate and spin functions. Figure 6-8 shows that the direction of motor rotation is controlled by cams B and D. For the agitate function, contacts 4 and 6 of cam B are closed, while contacts 5 and 10 of D are closed. During spin functions, contacts 5 and 6 of cam B are closed and cam D has contacts 4 and 10 closed. Because it has been determined that during the agitate function the motor starts properly, you may also assume that contacts 4 and 6 of cam B, and contacts 5 and 10 of cam D are operating properly. To check the timer output, examine contacts 5 and 6 of cam B and contacts 4 and 10 of cam D. This could be done with either voltage or resistance checks. To check voltage, place the timer dial at spin function of any cycle and measure the voltage. To check cam B, place one probe of the voltmeter on pin 18 and touch the other probe to pin 6. The reading should be approximately 120 Vac in this case. A similar check can be made of cam D by connecting one probe of the meter to pin C and touching pin 10 with the other. Again the reading should be about 120 Vac.

Before making any resistance measurement ensure all power has been removed. With timer in the spin position, remove the wires from pins 5 and 6 on cam B and measure the resistance between the pins. Also remove the wires from pins 4 and 10 and check the resistance between the pins. In both instances the reading should be zero ohms.

Keep in mind that other components could cause these same symptoms.

Motor Testing

Motor testing was described in detail in an earlier chapter of this book and will not be repeated here. Should you suspect motor failure, be sure to take the time to examine motor mountings and belts, as these can sometimes cause motor-shaft binding.

TROUBLESHOOTING

Electrical system troubles are located by observing the machine operation and comparing the actual operation with that specified in the timing sequence and schematic. Once you have become familiar with several models this process will be easy, and in many cases you may find you don't even need the timing diagram. The fact that the machine fails to perform a certain operation, or fails to sequence from one operation to another, will usually narrow the possible causes of failure. Electrical failures are usually easy to locate if you observe the machine operation closely.

Observing a complete wash cycle would take a considerable amount of time. This is done by most technicians only after other methods of troubleshooting have failed. A more popular method is to *short cycle* the machine. To do this, select the lowest possible water level and set the timer to the *wash/fill* position of the normal cycle. Start the machine and allow it to fill to the selected level. While the machine is filling, vary the water-temperature control and make sure that changing the control causes a corresponding change in water temperature. This is also a good time to discuss the problem with your customer. Once the machine has filled note that it automatically starts the wash/agitation cycle. If the water-level control is variable, turn the machine off and reset the water-level control to a higher setting. Restart the machine. It should once again start to fill. Again, the machine should stop filling and begin to agitate when the proper water level is reached.

If all is okay the timer can be advanced by hand to the end of the wash/agitate cycle, and the pumping and spinning operations can be observed. The machine is then allowed to fill for the rinse cycle and temperature check, as was done with the wash water. After this cycle has been determined to be operating correctly, the timer can be advanced to the spin phase. Observation time for the complete cycle of the electrical system is reduced to about one-third the time required for the machine to complete the wash cycle on its own.

Electrical problems will usually show themselves as a failure of the machine to complete the fill, agitate, pump, or spin operations. When a failure occurs, stop the machine and investigate further.

Troubleshooting

Automatic Washers

METHODS OF SERVICING THE THREE MAJOR SYSTEMS AND their components were discussed in previous chapters. Isolating the problem to a sufficiently small area is sometimes a very difficult task for the inexperienced technician. Troubleshooting charts, especially those that list an exact part failure for a particular symptom, do not always lead you in the right direction. Because of differences from make to make and model to model, troubleshooting charts are somewhat limited in scope. They are useful, however, to the extent that they give the troubleshooter ideas.

Included in this chapter is a series of troubleshooting charts designed to be useful with any automatic washer. You have seen from earlier discussions that all automatic washers perform the same functions and, in general, go through the same cycles of operation. The specifics of the job may vary, but the job itself is the same. For example, all machines have a wash/fill period, a wash/agitate period, and a rinse/agitate period. To broaden the application of the troubleshooting charts shown in this chapter, we have written them to guide the technician in analytical checks of machine operation rather than point him to a definite component failure. Manufacturer's service literature, especially timing charts and wiring schematics, should be used to locate the mechanism that accomplishes the function listed in these diagrams. Specific checks of some components will be recommended where they are common to many washers.

The troubleshooting charts are most useful when the technician observes the machine in operation and notes discrepancies. To do this, the technician attempts to operate the machine in a normal manner investigating malfunctions as soon as they are noticed.

WASH/FILL

All machines must go through a period in which the machine fills with water prior to washing the clothes. The purpose of the wash cycle is not only to fill the tube with water, but to insure that the water is at the correct level and temperature. If water level and temperature selection is permitted on the machine you are servicing, set the water-level control to low and water-temperature control to cold. Rotate the timer dial to the start of the normal wash cycle and start the machine. The trouble chart in Fig. 7-1 can be used as a checklist as you observe this operation.

Does the machine make any attempt to start the fill operation? Is there any indication that the machine is trying to start? If not, make sure the machine is connected to an electrical outlet and voltage is present at that outlet. If power is applied to the machine, vary the water-temperature control. Select hot water and notice if water enters the tub. Do the same with the water-level control. If varying either of these controls causes the machine to start filling, investigate the circuits associated with that control and the water inlet valve.

Should the controls have no effect, move the timer slightly. Sometimes the markings on the timer dial are not very accurate and you may be slightly off the *fill* position. If this does not reveal any clues, check hoses and faucets. Make sure they are properly connected and turned on. The next step is to check the water valve, as explained in an earlier chapter. After this, check the water-level control and water. As a last resort, check the timer circuits. All these checks should be made with the aid of a schematic and a VOM.

Once the machine starts to fill, check for leaks by looking under and behind. If you see any water, check hoses and seals. To perform this check, it may be necessary to move the machine away from the wall and remove the back panel. With water entering the tub and no visible leaks, vary the setting of the water-temperature selector and see if changing the water-temperature control brings about a corresponding change in water temperature. Should temperature problems be encountered, check for proper connection to the water supply, the temperature control circuits (including the inlet valve), and the timer, in that order.

If everything is operating satisfactorily at this point, the only remaining check is the water-level control. To check this control

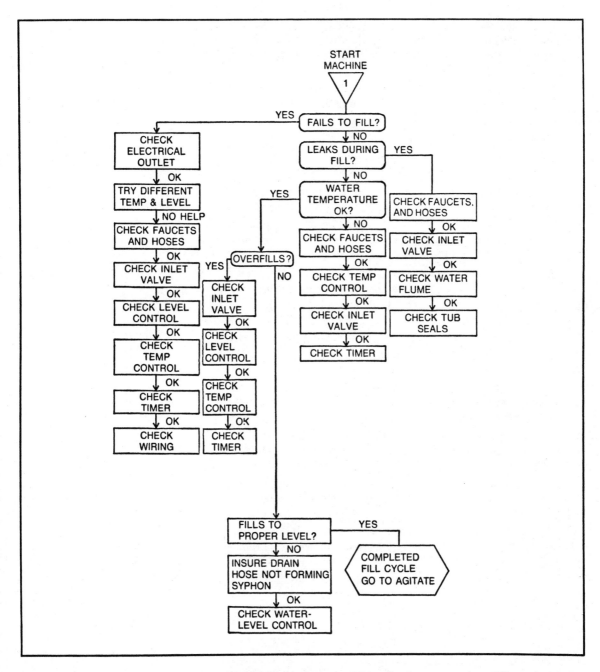

Fig. 7-1. Troubleshooting chart for the wash/fill cycle.

allow the machine to fill to the lowest setting. When the water reaches the prescribed level and is turned off, move the control to a higher setting. Check for water shutoff at several settings of the water-level control. Should the water fail to shut off and the machine overfills, unplug the machine. If this fails to stop the flow of water, turn off the water valve behind the machine. When unplugging the machine does not stop the water from entering, you will usually find a stuck inlet valve. If unplugging stops the water-flow, look for problems in the water-level control, water-temperature control, or timer circuits.

Once the machine fills with water to the correct depth and proper temperature, and turns off the incoming water, the wash/fill operation is complete and you are ready to observe the wash/agitate function.

WASH/AGITATE

The purpose of the wash/agitate operation is to flex the soiled clothing in the wash water. Agitation should be smooth and steady. In machines equipped with automatic dispensers, detergents are released into the water during this period. If the machine features a recirculating filter, lint filtering takes place throughout the wash/agitate operation. (See Fig. 7-2 for troubleshooting this cycle.)

The machine should start the wash/agitate operation immediately after filling with water. Should this fail to happen, the most likely suspect is the water-level control. This control signals the mechanical system to start the agitate period. During the wash/agitate operation, the water-level control should be in the *full* position. This control can be checked by voltage and resistance measurements as explained in an earlier chapter.

If the water-level control is functioning as it should, your next move will depend on what type of machine you are servicing. Some machines use a solenoid to shift the mechanical components to the agitate mode. Others use the timer or motor-switching circuits. A wiring diagram should be used to trace the agitate circuit and make necessary repairs.

When the machine shifts into the agitate mode but operation is noisy, check for worn belts or frozen pulleys in the mechanical system. Bearings and internal transmission parts can also cause noisy operation. Pinpointing these defects usually requires major disassembly.

Troubles in automatic dispensing mechanisms can usually be traced to that device or to the timer. Use a wiring diagram and follow the circuit for these components. Make sure any tubing used to route the laundry agent from the tank to the tub is not clogged.

Most machines have some sort of brake to hold the spin basket steady during agitation. The spin basket tends to rotate if not held by the brake. This will reduce the effectiveness of agitation. Should this failure occur in a machine you are servicing, check the solenoids associated with the brake mechanism. Look for trouble in the brake linings or springs if the solenoids are working.

If the machine has operated properly up to this point, look for

Fig. 7-2. Troubleshooting chart for the wash/agitate cycle.

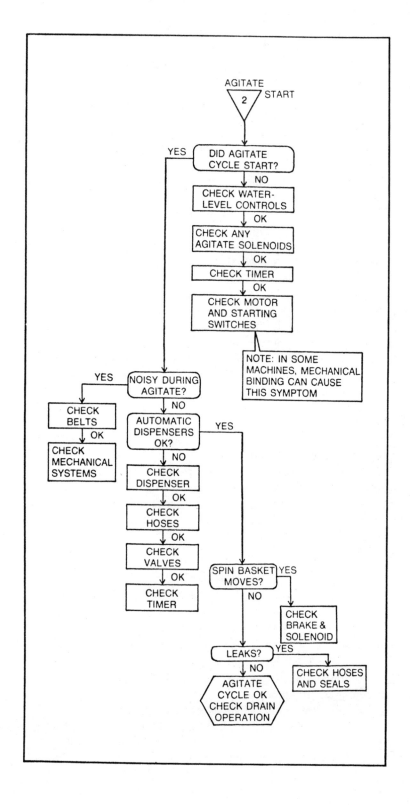

AGITATE

2 START

DID AGITATE CYCLE START? — YES
NO
CHECK WATER-LEVEL CONTROLS
OK
CHECK ANY AGITATE SOLENOIDS
OK
CHECK TIMER
OK
CHECK MOTOR AND STARTING SWITCHES

NOTE: IN SOME MACHINES, MECHANICAL BINDING CAN CAUSE THIS SYMPTOM

NOISY DURING AGITATE? — YES
CHECK BELTS
OK
CHECK MECHANICAL SYSTEMS

NO
AUTOMATIC DISPENSERS OK? — YES
NO
CHECK DISPENSER
OK
CHECK HOSES
OK
CHECK VALVES
OK
CHECK TIMER

SPIN BASKET MOVES? — YES
NO
CHECK BRAKE & SOLENOID

LEAKS? — YES
NO
CHECK HOSES AND SEALS

AGITATE CYCLE OK CHECK DRAIN OPERATION

leaks around tub seals and hoses. This would complete the check-out of the agitate operation.

WASH/DRAIN

The wash/drain operation varies somewhat from model to model. In some washers this function starts immediately upon completion of the wash/agitate cycle and continues until all water is removed. In other models, the water level recedes to about half capacity, then water is allowed to enter the tub to rinse soap scum. After this half-fill operation, all water is pumped from the machine. Another method pumps all water from the machine and refills it completely, after which a second pumpout is followed by the wash/spin period.

The wash/drain troubleshooting cycle begins with the completion of the wash/agitate period (Fig. 7-3). First note the action that occurs when the timer shows the wash/agitate time has ended. Compare this and subsequent actions to those shown in the timing diagram. If the machine fails to make any attempt to start pumping, check the drainhose for kinks or plugging and the lid switch for proper operation. Should it appear that the hoses are properly installed, check drive belts and the pump to make sure they are free to turn. The brake solenoid (on some models) and the water-level control can also cause the pump to fail. Sometimes you will find failures in the motor reversing circuits, motor-start switch, and the motor itself as the cause of pumping troubles.

Should the wash/drain operation start properly as outlined above, check tub seals and hose fittings for leaks while the water is draining. Completion of the drain period will be followed by a spin operation to remove the sudsy water from the clothing. If the machine starts the drain period, but fails to finish and enter the spin mode, check the lid switch, hoses, and water-level control, in that order.

WASH/SPIN

Following the pumping out of the wash water, the spinning operation starts to remove water remaining in the clothes. Most machines employ a spray/spin mode in which the clothing are revolved at high-speed while being sprayed with cool water. As shown in Fig. 7-4, the first things to check at this time are the starting of the operation, the amount of torque applied to the basket, operation of the lid switch, and the spraying (if used).

Failure to spin points to a check of the lid switch, water-level control, timer, and motor circuits. Mechanical failures in the transmission gearing or clutch system can also cause these symptoms. Check the solenoid-operated brake or spin solenoid on machines using these components.

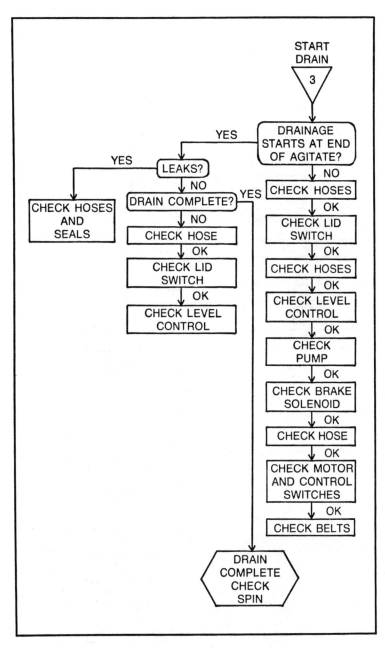

Fig. 7-3. Troubleshooting chart for the wash/drain cycle.

As spinning starts, notice the amount of torque applied. Revolutions should build fast with an empty tub. Notice the speed with which revolutions are added and the smoothness of the spin. Faulty bearings in the mechanical components of the spin mechanism can cause sluggish and noisy operation. When the basket is spinning at its normal rate, check the lid switch by opening and closing the lid. The spinning should stop with the lid open.

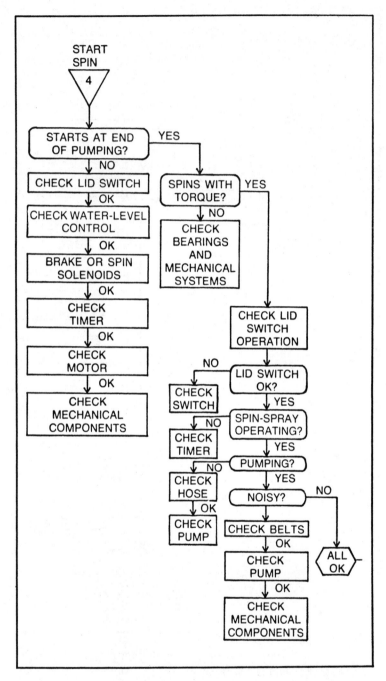

Fig. 7-4. Troubleshooting chart for the wash/spin cycle.

On machines using a spin/spray operation, note if spraying is taking place. Although lack of spray can be caused by a faulty waterlevel switch or water inlet valve (on some machines), the timer is a more likely suspect. The balance of the machine affects its stability. An unbalanced machine tends to "walk." Make sure the load is balanced and check machine leveling. If this doesn't locate the problem, look for loose mechanical parts.

Noisy operation can be caused by belts or mechanical components. Observation is the best way to locate these faults.

At the end of the spin period the machine will usually go through a short pause (about one minute) after which the rinse and fill operations are started.

RINSE

Troubleshooting the rinse, fill, agitate, and drain operations generally follows the same procedures used for the wash cycle. Keep in mind here that many symptoms seen in the rinse period would have been discovered in wash operation, because in most cases the same components are used. When troubles are encountered in this period, and these same troubles are not duplicated in the wash mode, use a schematic diagram to locate components not common to both cycles. If an electrical component that would cause such a problem cannot be located, check the manufacturer's service literature for mechanical components that are not common.

SPIN

The final spin period is no different from the previous spin period, except that it lasts longer to remove the water more thoroughly. One manufacturer recommends a weight test to determine if the proper amount of water is being removed. Weigh the clothes prior to loading and again after washing. The wet weight should be approximately twice the dry weight.

FUNCTIONAL TROUBLESHOOTING

Although used for some time in other industries, the concept of functional troubleshooting has only recently developed in home appliance repair. This method of fault isolation breaks a complex piece of equipment such as a washer into several operational areas without regard to whether the components in that area are mechanical or electrical. In this way, the total operation of very complex equipment, including the electrical-mechanical interface, can be checked much faster than if you first try to determine whether the problem is electrical or mechanical, then locate the particular component that is at fault.

For an example of functional troubleshooting applied to an automatic washer, refer back to the timing diagram in Fig. 6-6. This chart is divided across the top into several columns (fill, agitate, spin, etc.). Notice that, regardless of whether the machine is in the wash cycle or rinse cycle, the same repetitive actions occur. First, there is a period of time while the machine fills. An agitate period follows, then a spin period. The same cycle is repeated for the spin mode. The concept of functional troubleshooting consists of operating the machine through its various cycles and observing which function is not performing as it should. Outputs of that function are then checked to make certain the failure is occurring in that area. Inputs to that function are checked to insure that the fault is not caused by the input and is actually occurring in the area in question.

The following functional diagrams are generic and represent no particular make or model. The diagrams and their accompanying text are meant to give the reader an understanding of the functional troubleshooting process rather than the details of any one machine.

Power Function

As its name implies, the purpose of the power function, shown in Fig. 7-5, is to furnish the necessary electrical power to the other functional areas. In the machine represented in this text, the power function has three inputs. A mechanical input, representing the cycle selected, is received via the timer dial. By manually positioning the timer dial to a specific cycle the operator causes contacts that furnish outputs to the other functional areas to be closed. The second mechanical input is a signal from the timer itself that it has

Fig. 7-5. Power functional diagram.

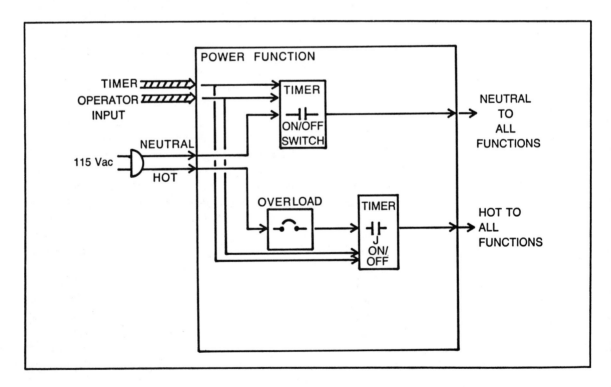

reached the end of the operating cycle. Electrical power is received by means of an appliance power cord plugged into a convenience outlet.

Three major components comprise the power function. A set of on/off contacts located in the timer connect the electrical components located in other functional areas to the power source. These contacts, operated by in and out motion of the timer dial, must be closed for any of the other functions to operate. Another set of contacts, operated by movement of cam J, connects the other side of the 115 Vac line to the remaining functions. Cam J is positioned by rotary motion of the timer shaft and therefore acts as an "off" signal when the timer reaches the end of operation position.

Overload protection is provided by the thermal overload device in the 115 Vac line to contacts J. If an over-current condition occurs, the heat generated will cause the contacts to open and remove power. Although this component is functionally located in the power area, its physical location could be anywhere in the machine. This device is usually located inside the motor frame near the windings. In some cases the overload will only remove power from the motor, while allowing other components to remain energized. Close examination of the manufacturer's wiring diagram will be necessary to determine these details.

Because the power function furnishes power to all electrical components in all the other functions, a failure in this function usually causes a complete loss of all operational functions.

Fill Function

The purpose of the fill function controls the filling of the tub with water. This function is always the first functional operation of either the wash or rinse operations and will always be followed by an agitate operation.

Electrical power for the fill operation is received from the power function, as shown in Fig. 7-6. In addition, two other operator-controlled inputs are received from the machine console. These inputs are mechanical and represent the desired water level and temperature, as selected by the operator. Only one output is available from the fill function. This output serves as a start signal to the agitate function.

Several signals are generated and used internally by the fill function. These signals control the operation of the fill function and have little or no effect on any other functions except for the start agitate signal.

When power is received from the power function, the fill function will immediately cause the machine to fill with water. Controls set by the operator determine the amount and temperature of the water.

The fill control compares the actual level of the water with the selected level. Control signals from this component allow the temperature selector to open the proper solenoids to fill the tub with the desired mix of hot and cold water. Because the desired temperature may change, depending on whether the machine is in a wash or a rinse operation, the timer furnishes signals to the temperature control that designate that either a wash/fill or a rinse/fill operation is about to begin.

Agitate Function

Upon completion of fill operation, a start agitate signal is sent from the fill function to the agitate function (Fig. 7-7). Electrical power is received from the power function and routed to the motor brake and timer.

The timer serves two purposes in the agitate function. First, it routes the electrical power to the motor in such a manner that the motor runs in the correct direction for agitate. Second, the timer furnishes the start spin signal at the end of agitate operation. The brake locks the spin basket to prevent it from moving as the dasher moves back and forth. Agitate mechanical power is furnished by the motor and transmission.

When the agitate function receives the start agitate signal from the fill function, the spin basket is locked and motor is started. The timer begins to sequence through the agitate time period. When the timer determines that the specified time period has passed, the motor and brake circuits are opened and the start spin signal is sent to the spin function.

Two key troubleshooting points are located in this function: the start signal which starts these events, and the start spin which signals the end of agitate operations.

Spin Function

Two things occur during the spin function. Water is extracted from the clothes, and that water is pumped out of the machine. This is achieved by pumping down the water level, then spinning the load at a high rate of speed while the pump continues to run.

The spin function, shown in Fig. 7-8, receives two inputs, one from the power function and one from the agitate function. Electrical power is routed through the lid switch to prevent the motor from starting during the spin function if the lid is closed. The timer reverses the inputs to the motor, causing it to run in the opposite direction. This reverse action energizes the pump and the spin basket. Electrical signals from the timer release the brake, allowing the spin basket and agitator to turn together.

Figure 7-9 illustrates the major steps of a typical functional

Fig. 7-6. Fill functional diagram.

145

Fig. 7-7. Agitate functional diagram.

Fig. 7-8. Spin functional diagram.

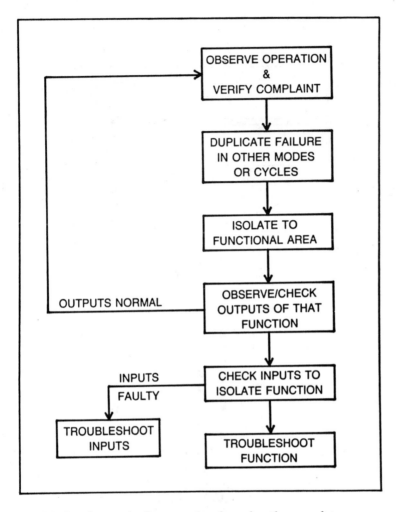

```
        ┌─────────────────┐
        │ OBSERVE OPERATION│
        │        &        │
        │ VERIFY COMPLAINT │
        └─────────────────┘
                 │
                 ▼
        ┌─────────────────┐
        │ DUPLICATE FAILURE│
        │  IN OTHER MODES  │
        │    OR CYCLES     │
        └─────────────────┘
                 │
                 ▼
        ┌─────────────────┐
        │    ISOLATE TO    │
        │  FUNCTIONAL AREA │
        └─────────────────┘
                 │
                 ▼
        ┌─────────────────┐
OUTPUTS │  OBSERVE/CHECK   │
NORMAL  │  OUTPUTS OF THAT │
        │    FUNCTION      │
        └─────────────────┘
                 │
                 ▼
        ┌─────────────────┐
INPUTS  │ CHECK INPUTS TO  │
FAULTY  │ ISOLATE FUNCTION │
        └─────────────────┘
                 │
┌──────────┐     ▼
│TROUBLESHOOT│  ┌─────────────────┐
│  INPUTS   │  │  TROUBLESHOOT    │
└──────────┘   │    FUNCTION      │
               └─────────────────┘
```

Fig. 7-9. Functional troubleshooting flow chart.

troubleshooting cycle. Start out by observing the complete operation cycle, keeping in mind the customer's complaint. Look for the same symptoms in other parts of the cycle. Vary the settings of the operator's controls and observe the effect on the symptom. This can sometimes be very helpful in pinpointing failures. If the symptom can not be duplicated, consult with the customer to eliminate the possibility of any misunderstanding. Repeat the cyclic tests several times to make sure you have duplicated the problem exactly. In most cases the malfunction can be isolated to a particular functional area using only the operator's controls, the customer's description, and observations of the operating cycles.

Once the malfunction is isolated to a particular function, electrical and mechanical tests should be made to verify that the failure is contained in that specific function. Aim these tests at the output of the function in question first. This will confirm that a failure does, in fact, exist in that area. Next, check the inputs to that

148

function to insure that the faulty functional area is not caused by inputs to the function.

If the outputs of a function are not correct and the inputs are operating properly, direct troubleshooting efforts at that function. Should the inputs prove to be faulty, repeat observations of the operating cycle to make sure your observations were correct. Another approach would be to troubleshoot the faulty inputs.

Gas Dryers

A NY MODERN CLOTHES DRYER, WHETHER POWERED BY GAS or electricity, uses three types of action to dry clothing: heat—applied by either a gas or electric heater to evaporate water in the clothing; air—applied by an electric blower or fan; and tumbling—applied by an electric motor to aid in air circulation and prevent burning the clothes. Notice that both tumbling and air circulation are accomplished by electrical devices, even in gas dryers. Gas dryers are those that use gas as a source of heat, as opposed to those using electrical sources.

GAS BURNERS

The source of heat for a gas dryer is a gas burner designed to operate on natural, manufactured, or liquid-petroleum (LP) gas. Early models required the operator to manually light the burner each time the dryer was used. This inconvenience was eliminated by using a pilot-light system to automatically light the burner when the dryer was turned on.

Pilot

The basic pilot light and safety valve combination is still found on many dryers. Its simplicity makes it ideal for starting our study of gas burners. Such a system is shown in Fig. 8-1. Most city plumb-

Fig. 8-1. A basic pilot and safety valve assembly.

ing codes require that each gas outlet be equipped with a manually operated shutoff valve outside the appliance. Gas enters through this valve and travels through piping to reach the pilot safety valve. The pilot safety valve has two outlets, one to the main burner and a smaller outlet to the pilot.

Both outlets are controlled by a thermocouple extending from the valve to the pilot flame. The job of the thermocouple is to sense whether the pilot is burning and shut off the gas whenever the pilot is out. A valve bypass is usually provided to manually light the pilot. This bypass is a pushbutton arrangement that allows gas to reach the pilot but not the burner. Pressing the button permits gas to flow to the pilot. Once lighted, the button must be held in until the thermocouple is hot enough to sustain the pilot. Once this happens the button can be released. With the pilot ignited, the burner will light any time gas is present.

A complete gas-burner control using this pilot system is shown in Fig. 8-2. Pilot operation is identical to that in the previous paragraph. A solenoid valve between the burner and pilot safety valve controls the electrical system. With the aid of the pilot and safety valve assembly, this valve allows the user to start the dryer from the control panel. The burner can then be completely hidden from view except when service panels are removed. Notice that the pilot gasline bypasses the solenoid control valve. The pilot remains lit when the dryer controls are in the *off* position.

With the pilot lit, the operator can load the dryer and set the timer to the *start* position. The drying process will begin. The timer sends an electrical signal to the burner control valve causing the solenoid to open the line to the burner. The pilot ignites the fuel as it reaches the burner, and the dryer is in operation.

This pilot system uses a pilot with a constant flame. That means the pilot is consuming fuel even when the dryer is not in use for long periods of time. To cut down on the amount of wasted fuel, a variation of this system was designed that uses a pilot with a vari-

Fig. 8-2. A complete burner system using a solenoid control valve.

able flame. When the dryer is not used the pilot burns at a low level. The flame doesn't reach the thermocouple and, therefore, doesn't open the safety valve to allow gas to the burner. This condition is shown in Fig. 8-3. When the operator turns on the dryer, the control system allows more gas to reach the pilot and the flame level increases. With the pilot flame now reaching the thermocouple, the safety valve will soon open and allow the burner to light.

This system has several disadvantages when used in automatic dryers, and was not in production for very long. It is still used successfully with gas ovens, however.

Combination Valve

One of the most popular burner control assemblies used with gas dryers is shown in Fig. 8-4. You will see this type of valve often if you service many automatic dryers. It has been in use for many years and is still found on some late-model machines. Although they may differ somewhat in size or shape, most combination valves work in the same manner.

Gas from the supply line enters the valve through the lower part of the regulator chamber. The purpose of the regulator is to

Fig. 8-3. A two-level pilot with a constant flame.

MERCURY BULB

BURNER PILOT ASSEMBLY

ADJUSTMENT SCREW (NATURAL & MANUFACTURED ONLY)

PILOT SWITCH DIAPHRAGM ASSEMBLY

LATCHING LEVER

LATCHING PIN

LATCHING CAP

PILOT VALVE SHAFT

MAIN BURNER VALVE SHAFT

COVER ASSEMBLY

MAIN BURNER COIL ASSEMBLY

SPRING

MAIN VALVE DISC

VENTED SCREW (NATURAL & MANUFACTURED ONLY)

TAILPIECE

GAS FLOW

UNION NUT

PILOT VALVE DISC

SAFE-LITE VALVE DISC

RESET SHAFT ASSEMBLY

PILOT VALVE SPRING

MAIN BURNER GAS PASSAGE

PILOT GAS PASSAGE TO FILTER

RESET SPRING

RESET BUTTON

Fig. 8-4. The combination valve is used in many gas powered dryers.

continuously supply gas at a constant pressure. This valve controls the amount of pressure that reaches the pilot valve in the upper chamber of the regulator stage.

To light the pilot on this unit it is necessary to push the reset button and hold a match to the pilot burner. The reset button opens the pilot valve and allows gas to enter the lower chamber of that section. The supply line for the pilot is tapped off the lower section of the pilot valve chamber. Note that the solenoid valve remains closed and seals the main burner outlet. When the thermocouple is heated by the pilot it exerts pressure on the pilot-switch diaphragm, engages the pilot latching cap, and holds the pilot valve open. The pilot will continue to burn when the reset button is released.

With gas pressure in the upper chamber of the solenoid valve, the dryer is ready for operation. When the operator turns on the dryer, the control system opens the solenoid valve and the pilot lights the burner.

Glow-Coil Pilot

A standing pilot wastes fuel when the dryer is not in operation. To overcome this, several electrically operated pilot and burner-control mechanisms have been marketed. One type uses a hot coil of wire to ignite the burner, another type uses electrical sparks. Both accomplish the same thing. No gas is consumed unless the dryer is in operation. Operation of a typical glow-coil control is explained with the aid of Fig. 8-5.

This system uses a pilot ignited by a wire that is heated with electric current. In Fig. 8-5 the pilot-control relay is actuated by the action of the thermocouple. When the dryer is idle the thermocouple relay is in the *cold* position. When the start switch is closed, current flows through the start switch, the warp switch (a thermostatic normally closed switch), and the pilot valve solenoid to the opposite side of the line. A second path exists through the resistor to the *cold* side of the thermocouple relay, through the cold contact, to the opposite side of the line.

Current in the primary of the transformer induces current in the secondary, and the glow coil starts to heat. The transformer is used to step down the 120 Vac to a lower level, usually above 24 V for safety. During this time, current through the pilot-valve solenoid causes that valve to open and supply gas to the pilot light.

Fig. 8-5. The glow-coil igniter consumes gas only when the dryer is in actual operation.

Heat in the glow coil builds rapidly and the pilot should light very quickly. When it does, the thermocouple relay senses the heat of the pilot flame and causes the center arm of the thermocouple relay to shift to the *hot* position. Current now flows in the main burner solenoid, opening that valve and supplying gas to the main burner, where it is ignited by the pilot.

The warp switch and resistor furnish a safety precaution for the burner control mechanism. Should the pilot fail to light within a short time, current through the resistor generates sufficient heat to cause the contacts of the warp switch to "warp" or bend open. This removes current from the glow-coil transformer and the pilot-valve solenoid, ensuring that the gas flow to the pilot is stopped. The thermocouple relay would stay in the *cold* position, and heat from the resistor would hold the warp switch open. If this happens the dryer must be turned off for about 10 minutes. This will allow the warp switch to cool before you attempt to relight the pilot.

Notice that the design of this system is built around the safety of the pilot. Both pilot and main-burner solenoids must be energized for the dryer to operate. Should the pilot fail to light, all power to the burner controls is removed. Glow-coil ignition systems not only offer this safety feature but save fuel as well because there is no standing pilot.

Spark-Ignition System

Another electrically controlled pilot uses an electric spark to light the burner, thus eliminating the pilot light completely. A schematic of the spark-ignition system is shown in Fig. 8-6. This system is typical of those using a burner directly ignited by electrical controls.

Assume the dryer has been standing idle and the electrical components are in the position shown in Fig. 8-6. The machine is loaded, the door closed, and the start switch pressed. This allows current to flow through the two switches to point A—the junction of several parallel circuits. One circuit is from point A through the safety-heater resistor, the coil of relay RY1, and the flame-sensor switch to point B. Current through RY1 causes the contacts of the relay to close.

Current also flows from point A through the contacts of the safety-heater switch to point C. Again the circuit parallels. One path goes through the igniter coil and the flame-sensor switch to point B. The vibrator is a small motor that opens and closes the spark contacts to produce an electric spark. The spark contacts are so positioned that the resulting sparks will ignite any gas present at the burner.

Another circuit exists from point C through the contacts of RY1 to the burner valve solenoid. This is paralleled by a path through

Fig. 8-6. The spark-igniter system eliminates the pilot light completely.

R1 to the burner valve solenoid. Current through the coil of RY1 causes the contacts of RY1 to close, shorting R1. This produces heavy current through the burner valve solenoid, opening the valve and supplying gas to the burner.

Burner ignition should take place immediately. Heat from the burner is detected by a heat sensor that controls the flame-sensor switch. The flame-sensor switch opens as soon as the burner ignites. This removes current from the safety-heater resistor and the igniter coil. However, current continues to flow through the contacts of the safety-heater switch. With no current through the coil of the RY1, the contacts are open and current flows through R1 and the burner valve solenoid. The resistor (R1) reduces the current that passed through the coil during the ignition phase. The burner valve is designed so that less current is required to hold the valve open than is required for initial opening. The purpose of this design will be explained shortly. First let's see what happens if the burner fails to ignite.

Should ignition not occur quickly after the start switch is closed, the safety-heater resistor causes the contacts of the safety-heater switch to open. This removes all current from the igniter coil and burner valve solenoid. This shuts off the gas and holds it off until the safety-heater resistor has time to cool.

Suppose the operator opens the door for a moment after ignition has occurred. Opening the door removes power from the burner valve solenoid. Gas to the burner is cut off. If the door is closed quickly enough, the only path for solenoid current is through R1 because the flame-sensor switch is still open. Because R1 doesn't allow enough current to open the burner valve solenoid, the system will not operate until the safety probe cools and closes the flame-sensor switch.

SERVICING AUTOMATIC BURNERS

Gas burners used in modern clothes dryers are reliable and can function for a long time without repairs. There are a few service steps that should be performed occasionally and, like any other electrical or mechanical device, repairs are sometimes needed. Cleaning and adjusting the airflow are the maintenance actions needed most often.

Over a period of time lint accumulates inside the dryer cabinet. Lint cannot only foul mechanical parts but it is also extremely flammable: So much so, in fact, that it is possible for large accumulations of lint to "flash" similar to gasoline vapors. This flash will usually occur during burner ignition and is sometimes followed by a less intense fire that burns and smolders in the lint. To prevent this, you should have a small vacuum cleaner in your toolkit and thoroughly clean all lint from inside the cabinet of every machine you service. This will not only help prevent fires but will reduce the need for servicing such failures as stuck centrifugal switches.

Adjusting the air supply to the burner is an easy service procedure. It will improve the efficiency of the dryer and reduce soot formation in the burner. Adjustment of air to the burner is accomplished as it is in other gas appliances. As shown in Fig. 8-7, gas enters the burner through a device that mixes the gas vapor with air. The flame should burn blue. Yellow flames indicate a dirty burner or an incorrect air setting. The air setting may be adjusted by loosening the setscrew and moving the damper to adjust the air inlet for clear, mostly-blue flames. Don't forget to tighten the setscrew.

Repairs to automatic burners depend on the type of pilot system used.

Standing Pilot

Failures, and likewise repairs, to the standing pilot are limited. In addition to cleaning and adjusting, failures of the pilot thermocou-

Fig. 8-7. Adjusting gas fires is accomplished by loosening the setscrew and moving the damper.

ple are common. Thermocouple failure is usually indicated by the pilot light not staying lit when the reset button is released. Complete failure of the pilot indicates a gas flow stoppage. Look for closed valves or plugged lines. If the pilot remains lit but the burner fails to ignite, check the main burner valve.

Glow Coil

Due to its complexity, the glow-coil ignition system is subject to more failures than the standing pilot systems. Failures can occur in either electrical or gas components. Checks for opens and shorts will reveal most electrical problems. Gas component failures are best located by observing the overall operation of the burner.

To troubleshoot burners rapidly, you can classify malfunctions into one of four groups: pilot does not light, glow coil does not light; pilot does not light, glow coil is hot; pilot light ignites but goes out; main burner fails to light with pilot burning.

Failure of the glow coil to light can be caused by a defective glow coil, defective transformer, defective warp switch, defective stop switch, or defective thermocouple relay. To check any of these components, remove all electrical power from the dryer. With an ohmmeter across the contacts of the start switch, press the start button (see Fig. 8-5). The meter should show zero resistance. Now check the contact resistance of the thermocouple relay in the same manner. Again the resistance should be zero. Once you have determined that these contacts are functioning properly, measure the resistance of the glow-coil (primary and secondary) transformer. If there are no opens or shorts check the wiring between the components. Although the design of glow-coil ignition systems varies from make to make, these checks should reveal any fault associated with a glow coil that will not heat. If you have made these checks correctly and still fail to pin down the fault, refer to the manufacturer's wiring diagram and check the electrical-control circuitry to the burner.

If the glow coil gets hot but the pilot fails to produce a flame, look for trouble in the area of the pilot burner. Check the continuity of the pilot solenoid. If the solenoid coil is good, restore power to the dryer and attempt to start the burner. Pay particular attention to whether you can hear the pilot valve operate. Should the pilot valve be operating properly, check for a clogged pilot line or gas-line to the dryer. A valve may be closed.

Should the pilot ignite but fail to sustain a flame long enough to cause burner ignition, look for problems in the pilot thermocouple, the thermocouple relay, or the solenoid valve. This problem is sometimes caused by air drafts. Therefore, make sure all burner covers are in place and that no lint accumulates around the pilot.

Suppose the pilot functions as it should but the main burner fails to light? Refer once again to Fig. 8-5. Note the components in the main burner circuits. A defective thermocouple, thermocouple relay, or main burner solenoid could cause these problems. Check continuity of the relay contacts and the main burner solenoid.

Spark Ignition

Visual inspection, close observation of operation, and resistance checks are the surest ways to locate the faulty component. Keep in mind the operating sequence for the spark ignition described in conjunction with Fig. 8-6. Try to relate the failure to the function of one of the components in the figure. Divide the failure into one of three categories: burner fails to light, igniter fails to operate; burner fails to light, igniter operates correctly; burner lights but will not stay lit.

Complete failure of both the burner and igniter could be caused by troubles in either the electrical control system or the burner assembly itself. Before attempting to repair the burner assembly, make sure you have isolated the problem to that area. In a later discussion we will explain the troubleshooting procedures used in locating problems in the electrical control system. For now, assume we have already isolated the problem to the burner assembly and need to pinpoint the faulty component. Components that can cause a complete failure of the main burner and igniter include the flame-sensor switch, the safety-heater switch, the igniter coil and contacts, and RY1.

In cases of igniter failure always check the operation of the igniter coil and contacts. Usually, you can hear the igniter immediately after starting the dryer. Sparking of the contacts makes a distinctive clicking sound. Check for lint accumulated around the contacts and for burned or pitted contacts. The manufacturer's service literature will give the proper setting for the spark contacts.

If servicing the contacts themselves will not pinpoint the problem, resistance checks of associated components should be made. Check for continuity across the contacts of the flame-sensor switch and the safety-heater switch. If either switch is open during the startup period, power cannot reach the igniter. Resistance readings should also be taken across the igniter coil, the coil of RY1, and the safety-heater resistor.

Should these tests fail to locate the faulty component, use the wiring diagram for the machine and check the burner control circuitry.

The burner should ignite with the igniter operating properly. If not, look for problems in the burner valve solenoid, the contacts of RY1, and R1. Continuity checks across the relay contacts and resistance measurements of R1 and the solenoid coil should reveal the problem.

If the burner light fails to remain lit, look for trouble in and around the burner as well as RY1, R1, and the solenoid coil. Visual inspection of the burner should locate any problems there. Resistance checks should pinpoint other problem areas.

GAS DRYER SYSTEMS

The gas dryer uses electrical power to operate the drive motor and control the sequence of events that occur during the drying cycle. Many of the components, such as timers and solenoids, are similar to those used in automatic washers. You will find, however, that timers and associated components used in dryers are usually simpler than those used in automatic washers. You will also encounter a new device—the thermostat. This is a very simple device used to control electric current. It is made of heat-sensitive bimetal that will open or close an electric circuit when heated, due to bending of the bimetal. In other words, it is a heat-sensitive switch.

Timer

Timers used with automatic dryers are built very much like those used in washers except that dryer timers have fewer devices to control and, therefore, have fewer contacts. For example, it is possible for an automatic dryer to be controlled by a timer with only three contacts. Figure 8-8 illustrates the wiring diagram of a timer.

Power from the 120 Vac source is received by the timer through the door switch. Cam A opens and closes the contacts to the timer motor. This motor drives the timer shaft, causing the other contacts to open and close at the proper time. Cam A is closed when-

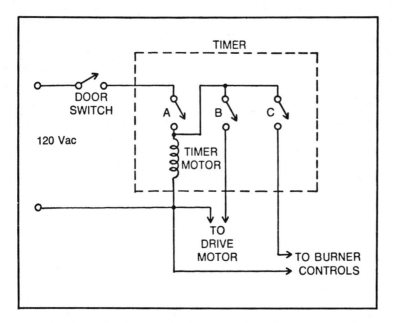

ever the timer dial is positioned at any point other than *off*. Cams B and C are designed to close the burner contacts after the motor contacts have been closed. Cam B supplies motor power and cam C furnishes power to the burner controls.

Fig. 8-8. A simplified wiring diagram for a dryer timer.

To operate a dryer with this type of control, the owner loads the machine, closes the door, and sets the timer to the desired drying time. The timer motor is energized and starts to drive the timer shaft. At the end of the drying time, the timer opens the contacts of cam A. This removes power from the motor and burner circuitry and completes the drying cycle.

Timed Drying Cycle

The drying cycle of a typical dryer is shown in Fig. 8-9. This graph illustrates the temperature inside the drum at various times in a drying cycle of approximately one hour. Principles explained here apply to both gas and electric dryers. To understand this chart, assume the dryer has been off for some time and the ambient temperature inside the drum is 80 degrees Fahrenheit. At time *zero* the dryer is loaded with a bundle of wet clothes. This causes a drop in the drum temperature, as shown on the graph.

With the door closed and the machine started, the temperature inside the drum starts to rise. At about 120 degrees Fahrenheit, moisture in the clothes starts to evaporate as hot air is circulated. The burner (in gas dryers) continues to supply a constant heat at a much higher level than 120 degrees Fahrenheit. However, evaporation of moisture in the clothing causes a cooling effect. This phenomenon will result in the drum temperature "flattening out" at some level and remaining at that level until most of the moisture has been evaporated. The temperature remains fairly constant during most of the drying cycle.

Near the end of the cycle, when most of the moisture has been removed, the temperature starts to rise quite rapidly. This is due to the lack of evaporation. At some point (usually around 160 degrees Fahrenheit) the temperature reaches a level that causes the contacts of a bimetal thermostat to open. This thermostat is in the circuit to the heating device. Power is removed from the burner solenoid in gas dryers. With the burner off, the temperature inside the drum begins to drop. If the clothes were removed at this time they would still be wet.

Although the burner was extinguished, the blower and drum continue to function. They tumble the clothes and circulate the air. As the temperature continues to drop, the thermostat cools and the contacts close. The burner is relit and the temperature rises to open the thermostat. This process continues until the timer cycles to the *off* position or the clothes are removed from the dryer.

Most fabrics contain moisture amounting to about 5 percent of their weight. Ideally, the timer should be set to a time that would dry the clothing to that level of moisture. This also means that the clothes in any one load should be about the same weight. If not, some might overdry while others underdry.

The drying cycle described here is known as a *timed cycle*. A timed cycle is one that runs for a selected time and stops. The dry-

Fig. 8-9. The timed drying cycle.

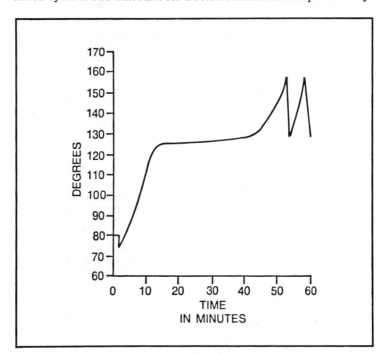

ing process is dependent on the time the load has been in the dryer rather than the actual moisture content of the clothing.

Automatic Drying Cycle/Auxiliary Thermostat

Many older dryers, although called automatic, feature only the timed-dry sequence. The development of synthetic fabrics led to the need for different drying cycles for different materials. This could be overcome, to some extent, by selecting different drying times. This however would not be truly automatic. Several methods have been used that employ a variable drying time. The operator may select one of two or more drying cycles to conform to the type of fabric being dried.

Figure 8-10 shows the control circuitry used with one automatic dryer. This system makes use of two thermostats to control the time the burner is on. The cycling thermostat is actually a heat-sensitive single-pole, double-throw (SPDT) switch. Physically, this thermostat is located where it is exposed to air expelled from the drum. Electrically, the contacts of the thermostat are wired so contacts 1 and 3 supply current to the burner controls while contacts 1 and 2 supply current to the resistor network. The cycling thermostat is designed to close contacts 1 and 3 and open contacts 1 and 2 at temperatures below 120 degrees Fahrenheit. At 130 degrees Fahrenheit contacts 1 and 3 open and contacts 1 and 2 close.

A second thermostat, sometimes called an auxiliary or bias thermostat, is also used in the control circuitry. This thermostat has its own heat supply in the form of a resistive heater. The contacts

Fig. 8-10. An automatic drying control using an auxiliary thermostat.

are located in the voltage supply line to the cycling thermostat. Current through the resistive heater controls the heat applied to the thermostat and, therefore, controls the opening and closing of the contacts. On this particular model, three different sizes of resistors can be alternately placed in series with the thermostat heater by means of switch contacts—one each for regular, wash-and-wear, and delicate fabrics. The smaller the series resistance, the more current allowed to flow through the resistive heater, causing it to heat faster. Current through the heater is controlled in such a manner that the contacts of the auxiliary thermostat open when the clothes are dry.

When the dryer is loaded, the operator selects one of the three automatic cycles (depending on the type of clothing) and starts the machine. Contacts 1 and 3 of the cycling thermostat are closed as are the contacts of the auxiliary thermostat. The burner is fired—generating heat for the drying process. No current is flowing through the resistive heater since contacts 1 and 2 of the cycling thermostat are open.

When the exhaust temperature reaches 130 degrees Fahrenheit, contacts 1 and 2 of the cycling thermostat close and contacts 1 and 3 open. This shuts off the gas supply to the burner and current flows through the resistive heater. With the heat source removed, the exhaust temperature drops. The temperature of the auxiliary thermostat increases due to current through the resistive heater. At a point where the temperature of the drum exhaust drops to 120 degrees Fahrenheit, the contacts of the cycling thermostat reverse—closing 1 and 3 and opening 1 and 2.

The burner is started and heat is applied to the drum. Since current has been removed from the resistive heater, heating of that unit ceases. The controls remain in this state until the drum exhaust reaches 130 degrees Fahrenheit and the contacts of the cycling thermostat are reversed—causing the auxiliary thermostat to heat. Between periods of current flow the auxiliary thermostat does not have time to dissipate very much heat. Therefore, each time current flows through the heater it is at a somewhat higher temperature than it was at the beginning of the previous period.

This cycle alternately applies heat to the drum and auxiliary thermostat. Each cycle is shorter due to residual heat in the drum. At some point, the auxiliary thermostat will reach a sufficient level to open its contacts. When this happens all power to the control circuit is removed and the drying process is ended. The resistance of the series resistors is determined by the manufacturer to ensure proper drying time for selected fabrics.

Automatic Drying Cycle/Cycling Thermostat

A variation of the control circuit for an automatic drying cycle is shown in Fig. 8-11. Operation of these controls depends on the switching of an SPDT thermostat. Upon initial loading, the machine is in a cooled state and contacts 1 and 2 of the cycling thermostat are closed. When the timer is placed to the automatic-dry cycle, timer contacts A and B are closed. This starts the timer. Current through contacts 1 and 2 allows the burner to start.

After about 15 minutes, timer contacts A and B open. Power is removed from the timer motor and the timer ceases to advance. The burner and main drive motor continue to heat and dry the damp clothes. When the temperature starts to peak (as shown in Fig. 8-9), the cycling thermostat opens contacts 1 and 2 and closes contacts 1 and 3. Removing power from the burner extinguishes the flame. The motor circuit remains closed and tumbles the clothing, however. With contacts 1 and 3 of the thermostat closed, the timer motor is started and the timer starts to advance. This phase of operation (burner off, timer running, and main motor running) continues for about 5 or 10 minutes. This is to allow the clothes to cool while tumbling to prevent wrinkling. At the end of this period the timer will cut off all power and stop the cycle.

ELECTRICAL SYSTEMS

A complete wiring diagram of a typical automatic gas dryer is shown in Fig. 8-12. This machine represents a "top of the line" model with several convenience features. It has two automatic cycles and one timed cycle. Four heat selections are available: low, high, no heat, and automatic. In the *no heat* mode the burner is not energized but the motor circuits are. This allows the clothing to be tumbled or fluffed without applying heat. Selection of either automatic cycle, one for regular fabrics and one for permanent-press

Fig. 8-11. An automatic drying control using a cycling thermostat.

Fig. 8-12. A wiring diagram for a typical automatic gas dryer.

fabrics, will automatically select the *high* mode. Another feature of this machine is called a *stop and dry* mode. In this operation, heat is applied but the clothes are not tumbled. Instead, a drying rack is placed in the drum and very delicate fabrics are placed on the rack. Operation in the stop and dry mode applies warm air to the load.

Timed Drying Cycle

In Fig. 8-12, 120 Vac is applied to the dryer between points A and B. The *B* line will serve as a common throughout the machine. The SPDT door switch serves two functions. When the door is open, the normally closed (NC) contact is made and the drum light is on. At the same time, the normally open (NO) contact is open, removing power from the rest of the machine. With the door closed, power is available at contact 1 of the timer and at the push-to-start (momentary-contact) switch.

Notice how the timer contacts are used to sustain motor operation once it has been started. Whenever the timer dial is moved from the *off* position, contacts 1 and 3 close, supplying power to the centrifugal start switch at contact 2. This switch remains open, however, until the motor reaches its proper operating speed. Therefore, contacts 1 and 3 of the timer cannot furnish starting power. Built into this timer is a momentary-contact switch on the timer shaft. Once the timer has been moved from the *off* position, pushing the timer shaft closes the momentary-contact switch. Power is supplied through the momentary-contact switch and contacts 1 and 3 of the motor centrifugal switch. (The function of the no-tumble switch will be discussed later.) Once the motor reaches operating speed, contacts 1 and 3 of the motor centrifugal switch open and contacts 1 and 2 close. The momentary-contact switch can now be released and the motor will continue to run. Contacts 1 and 3 will remain closed until the timer motor drives the timer switch to the *off* position.

Contacts 1 and 2 of the timer close at the same time contacts 1 and 3 close. Contacts 1 and 2 will open about five minutes before the end of the drying cycle. From contact 3, a jumper supplies power to the switching arrangement at contacts 4, 5, and 6. During timed-dry operations, contacts 4 and 6 remain closed until the end of the drying period. When contacts 4 and 6 open, contacts 4 and 5 close and a buzzer notifies the operator that the drying cycle has been completed.

Heat selection during the timed drying cycle is achieved by the heat-selector switch. In the *low* position, contacts 1 and 2 are closed, routing power through the low-heat thermostat to a set of contacts on the motor centrifugal switch. *High* heat is selected when contacts 1 and 3 are closed. Current for the burner unit is then through the control thermostat, motor centrifugal switch, and overload. When *timed-dry* is selected, timer contacts 4 and 6 remain closed during the entire operation. Therefore, the timer motor continues to drive regardless of the position of the contacts on the control thermostat. In timed-dry operations using high heat, the control thermostat will cycle the burner on and off but has no effect on the operation of the timer.

Automatic Dry/Regular Fabrics

To dry regular fabrics using the automated features of this machine, the cycle for automatic dry/regular fabrics is selected. Contacts 1 and 2 of the timer, the push-to-start contacts, and the motor centrifugal switch function just as they did in the timed-dry cycle.

Contacts 1 and 2 will close when the timer is positioned away from the *off* position. In this cycle these contacts are opened by the timer cams shortly before the cycle ends.

In automatic operation the heat-selector switch is set to the *high* position, eliminating the low-heat thermostat from the circuit and selecting the control thermostat. Burner-control current is supplied through contacts 1 and 2 of the timer, the high-heat contacts of the heat-selector switch (1 and 3), contacts 1 and 3 of the control thermostat, the motor centrifugal switch, and the overload. Notice that the overload is always in the circuit for burner-control current regardless of which cycle is selected. The overload limits the amount of heat generated by the burner, opening the circuit and removing power any time the temperature rises to a specified maximum. This prevents overheating and damage to the machine or clothing.

The most obvious difference between the timed-dry cycle and the automatic-dry cycle is the timing of the opening of contacts 4 and 6 in the timer motor circuit. When the regular-fabric automatic cycle is selected, these contacts close at the start of the cycle but open after a short period of operation. This removes current from the timer motor and stops the timer. The other operations of tumbling and heat application continue. When the temperature in the drum rises to a point indicating most of the moisture has been evaporated, the control on the cycling thermostat opens at contacts 1 and 3 and closes at contacts 1 and 2. As a result, the burner is extinguished and the timer motor starts while the main motor and blower continue to operate. Temperature in the drum will begin to drop after a short time and contacts 1 and 2 will open while 1 and 3 close.

At this time, the timer is stopped and heat is applied to the drum. When the drum temperature reaches the specified level, the control-thermostat contacts shift positions—the burner goes out and the timer comes on. This off-again, on-again operation continues until the timer advances to where the cam closes contacts 4 and 6 in the timer motor circuit. Operation from here to the end of the cycle is the same as it was for the timed-dry cycle.

Automatic Cycle/Delicate or Permanent-Press Fabrics

Operation of this cycle is similar to that used for automatic drying of regular fabrics. The major difference is the operation of contacts 4 and 6 in the timer motor circuit. In the regular fabric cycle, note the differences shown in Fig. 8-12. Since these contacts are opened at a point near the end of the cycle, during permanent-press cycles less heat is applied to the clothing in the last few minutes of drying. This reduces wrinkling by tumbling the clothing under low-heat conditions.

Fluff Cycle

This cycle allows the main motor and timer circuits to operate while disabling the burner-control circuits. The result is clothing tumbled in an air stream, but no heat is supplied to that air. Some housewives use this cycle to "finish dry" permanent-press fabrics or remove lint that clings to some synthetic fabrics when dried under high-heat conditions.

During this cycle, the contacts of the heat-selector switch remain open to prevent current from reaching the burner. Timer motor and drive motor circuits function the same as they do for timed-dry operations.

No-Tumble Dry

Some manufacturers have designed machines to meet the needs of those who prefer a drying operation that applies heat to the load without tumble action. This cycle is useful for items such as knit sweaters that might stretch out of shape in a tumble dry cycle.

To accomplish this, a special clutch is used on one end of a double-shafted motor. Figure 8-13 illustrates how this motor is connected to the drum and fan. One end of the shaft is equipped with a standard V-belt pulley attached to the shaft by a setscrew. The V-belt is used to drive the fan any time the motor is running. The opposite motor shaft is equipped with a special ratchet-type clutch that allows it to freewheel when the motor is turning in the reverse direction.

For example, when the motor is turning clockwise, both pulleys are locked to the motor shaft and both the fan and drum are driven. If the motor turns counterclockwise, the fan will still turn. However, inside the drum pulley is a ratchet mechanism that allows the motor shaft to slip inside the drum pulley. When the motor is turning in this direction, the drum pulley does not turn and the drum remains stationary. In using this mode, a rack is placed inside the drum and the article to be dried is placed on the rack. Other dryer controls operate as they would for any other cycle.

Figure 8-12 illustrates the wiring necessary to accomplish this type of operation. In the start circuit for the main drive motor is a switch labeled NO TUMBLE. For normal dryer operation the DPDT switch is made from 1 to 3 and 2 to 5. This establishes the normal polarity of the start winding in relation to the run winding. It results in normal motor rotation. Placing the switch in the *no tumble* position connects contacts 1 and 4 and contacts 2 and 6. This reverses the polarity of the start switch, and the motor will now turn backwards. Since the clutch slips, the drum cannot turn.

Fig. 8-13. A typical motor and pulley arrangement used for no-tumble operation.

Troubleshooting The Electrical System

Voltage and resistance measurements reveal the source of most troubles that develop in the electrical system. When taking voltage measurements, remember that power is being applied to the circuits. Take precautions to avoid shocks. If resistance readings are to be made, disconnect all electrical power.

AIR CIRCULATION

In order to dry clothing properly without overheating, adequate air circulation is required. A fan capable of moving several cubic feet of air per minute is used to move air through an arrangement of vents and ducts. Although the details of the air supply system vary from dryer to dryer, the overall operation is the same. Room air is drawn into the machine, heated, passed over the clothing, and expelled. Figure 8-14 is typical of these systems.

A large fan or blower is placed in the exhaust vent and pushes air out of the machine through the vent ducts. A low-pressure area is created in front of the blower. The draft of this low-pressure area pulls air from the room into the cabinet and through the heater. The tendency of hot air to rise and the action of the blower force the hot air up through the interior venting of the machine and through the vent holes in the drum. Once inside the drum, hot air picks up moisture and lint before being drawn into the lint trap, into the blower, and out of the machine.

Although simple in design, the lint trap plays an important part in the drying process. Lint picked up in the drum is very flammable. If not removed, it could collect in the blower and exhaust vent and result in a serious fire. Also, it can block the vent and prevent the passage of air, resulting in poor drying efficiency. Always check the condition of the lint trap when servicing automatic dryers. In-

INLET
AIR

EXHAUST
AIR

30%

70%

OUTSIDE AIR ➡

FLUE PRODUCTS ➡

struct your customers on proper cleaning procedures. Most manufacturers recommend cleaning the lint filter after each load.

To operate properly, the air circulation system requires that all gaskets and seals be in good condition and all ducts be free of obstructions. Anything that restricts the airflow within the machine will reduce drying efficiency. Likewise, a leaky gasket or seal can let air enter the system at some point other than the burner. This air, at room temperature, also reduces drying efficiency.

Fig. 8-14. A typical air circulation system used in gas dryers.

Electric Dryers

T HE GENERAL OPERATING CHARACTERISTICS OF AN ELEC-
tric dryer are not very different from those of a gas dryer.
The biggest difference is in generating heat. You will probably find
the electric dryer much easier to understand than the complex gas
dryer. Although some electric dryers can operate on standard
120 V house current, most (including those discussed here) operate on 240 V. Electric dryers operating on 102 V power are not
as efficient as their 240 V counterparts. The lower voltage causes
more power to be used and requires a longer drying time. These
disadvantages have decreased the popularity of these machines;
therefore, we limit our discussion here to those machines using 240
V power.

ELECTRIC BURNER

The electric dryer accomplishes heat generation with a high resistance wire-type heating element. The wire is formed in a small coil
(similar to a door spring) and can be strung in a large circle behind
the dryer drum or formed into a unit that fits inside the dryer in-
take ducts. Operation is the same. Only the location is different.

The element may be a single or double strand between
3000—6000 W. The double strand elements are used on machines
featuring two or three temperature settings where these settings

are controlled by the amount of current drawn by the element. Temperature control can also be achieved by a thermostat.

Single Element/Timed Dry

The circuit for a single element dryer offering only a timed-dry cycle is shown in Fig. 9-1. Power is supplied at points L1 and L2 at 240 Vac. The current to the element must flow through the timer contacts and safety thermostat before reaching the heating element. To prevent heat damage to the clothes, the safety thermostat will open the element circuit in case of overtemperature. The timer contacts stop the current flow at the end of the cycle. A centrifugal switch protects the clothing in case of motor failure. Without this device, should the dryer be turned on and the motor fail to start, the clothing and machine could be damaged by the high temperatures. The door switch will open the current path whenever the dryer door is opened.

A variation of this circuit is shown in Fig. 9-2. This dryer uses two thermostats and a single heating element to obtain three cycles of operation, all of which are timed. No automatic control of the drying time is employed. Instead, each cycle runs for a time determined by the timer setting. The timer contacts control the timer motor and heating element. Contact 1 supplies current to the timer motor whenever the timer dial is moved from the *off* position. Contact 2 furnishes heating-element current. The cam for contact 2 is cut so that during the normal fabric's cycle the timer will open contact 2 about three minutes before contact 1. During the permanent-press cycle, contact 2 opens about 5 minutes before contact 1. In both cases, current to the element is interrupted prior

Fig. 9-1. A simplified wiring diagram for a single-element timed-dry operation.

to motor cutoff. This prevents "baked in" wrinkles. Tumbling for a short period in the warm air tends to shake out wrinkles left by washing and drying.

As mentioned previously, this is a three-cycle machine. The heat-selector switch and the two heat-control thermostats make this possible. When high heat is selected, contacts 1 and 3 of the heat-selector switch are closed; low heat closes 1 and 2. Contacts 2 and 3 are open during the *fluff dry* cycle, removing all current to the heater element. The high-temperature thermostat opens at 145 degrees Fahrenheit and closes at 135 degrees Fahrenheit. The low-temperature thermostat opens at 135 degrees and closes at 125 degrees. The safety thermostat and the centrifugal switch in this example operate in the same manner discussed in conjunction with Fig. 9-1.

Single Coil/Automatic Dry

Figure 9-3 shows the heating-element control circuit for a single-coil dryer with automatic control. In addition to the regular timed cycle, this circuit features two automatic drying cycles. One is for regular fabrics and the other is for permanent-press fabrics. The heat-selection options found in the previous control system are also found in this system.

The only difference in the two automatic cycles is the timing for opening timer-contact 2, which supplies power to the timer motor. Timer-contact 2 opens about 3 minutes before the end of the drying cycle for regular fabrics and about 5 minutes before the end of the drying cycle for permanent-press fabrics. This removes power from the heating circuit and allows time for the clothes to cool for easier handling.

The timer motor opens contact 2 about 15 minutes before the

end of the cycle to achieve automatic operation. The automatic-control thermostat now controls both the heating element and the timer motor. Contacts 1 and 3 of the automatic-control thermostat close when the temperature falls below a preset level. Current for the heating element is now supplied through these contacts and the safety thermostat.

As the temperature rises, contacts 1 and 3 in the automatic-control thermostat open and contacts 1 and 2 close. This removes current from the heating element and supplies it to the timer motor. The timer advances and the drum temperature drops. When the temperature is low enough, the automatic-control thermostat is reset and the heating element raises the drum temperature. This sequence continues until the timer advances to the *stop* position and opens timer-contacts 1 and 2.

Dual-Element Heater

Figure 9-4 is a wiring diagram for the dual-element heating circuit. This element offers a closer control of drum temperature by applying a more constant supply of heat.

In the timed cycle, timer-contacts 1 and 2 remain closed. This supplies current from L1 of the 240 V supply to the timer motor. Current for the heating element is supplied between L1 and L2 and passes through the safety thermostat and the centrifugal switch.

To achieve automatic operation, timer-contacts 1 and 2 remain closed for the first 15 minutes of the drying cycle. When these contacts open, the heating element is controlled by the exhaust thermostat, and the timer is controlled by the control thermostat. These

Fig. 9-3. The heating-element control circuit for a single coil dryer with automatic control.

two thermostats open at about the same temperature. The control thermostat opens the timer motor circuit. The exhaust thermostat opens the circuit for the 3000 W heating element. This circuit differs from the single-element circuit in that it continues to supply heat when the timer motor is energized. This heat passes through the 2000 W heating element and is therefore somewhat lower. Other than this difference, the single-element and dual-element circuits operate the same.

ELECTRIC DRYER WIRING

Although there are many similarities, there are also major differences between gas-dryer wiring and electric-dryer wiring. The differences arise from the fact that electric dryers operate on a three-wire 240 V line, whereas gas dryers operate on the conventional 120 V line.

At this time you should review the chapter covering the service entrance and the 240 V service. In the next chapter we will look closer at the 240 V dryer hookup, but at this point we want to concentrate our attention on the circuits within the dryer itself. To do this, two dryers will be used as examples. One features only timed-dry cycles. The second has automatic dry and several other features as well.

Fig. 9-4. The wiring diagram for the dual-element heating circuit.

Timed Dry

Electric power is furnished to the dryer in Fig. 9-5 through a three-wire hookup. Between lines L1 and L2, 240 Vac is supplied. Between the neutral and either L1 or L2 will be 120 Vac. Close examination of Fig. 9-5 will reveal that the drive motor and timer motor circuits are supplied with 120 V since they are connected between L1 and neutral. The heating element circuit takes 240 V, between L1 and L2.

This model offers two cycles with three heat selections. When the heat-selector switch is set at *low*, contacts 1 and 2 are closed and the low-temperature thermostat is in the circuit. If *high* is selected, contacts 1 and 3 of the heat-selector switch are closed and the high-temperature thermostat is in control. For *fluff dry*, both contacts are open and no heat is supplied. The safety thermostat protects against overheating in all cycles. The centrifugal switch prevents heat from being applied until the motor is up to its proper operating speed. Contact 3 of the timer is closed when the timer is moved from the *off* position. It remains closed until about 3

Fig. 9-5. Wiring diagram for a timed-dry machine.

Fig. 9-6. Wiring diagram for a dryer featuring two automatic cycles.

minutes before the regular cycle is completed. If the permanent-press cycle is selected, timer-contact 3 opens about 5 minutes before motor cutoff.

Timer-contact 1 is a momentary-contact switch used to start the timer and main drive motors. Once the motors start, current is furnished through timer-contact 2 and contacts 3 to 5 of the motor's centrifugal switch. This circuit sustains current for the drive motor and timer motor throughout both cycles.

Automatic Dry

The wiring diagram of a dryer featuring two automatic drying cycles is shown in Fig. 9-6. One timed drying cycle and three heat levels are provided. This machine also has a *dry without tumble* feature. Essentially, this machine is the same as shown in Fig. 8-12, with two exceptions: the electric heating system, and the manner in which 120 Vac for the control circuit is tapped from the 240 V supply. Tracing the circuit in the diagram will illustrate these differences.

Between L1 and neutral, 120 Vac is routed to the main drive motor and control circuits. Voltage is supplied from L1 directly to one side of the drum light. The other side of the drum light is connected to a contact on the SPDT door switch. When the door is open, this contact closes and current flows through the light. Closing the door removes power from the light and connects the motor circuit between L1 and neutral.

Two timer contacts are in the motor control circuit. Timer-contact 2 is closed when the timer dial is set to any position other than *off*. This places 120 V (to neutral) at contact 3 of the motor centrifugal switch. Pressing timer-contact 1 (momentary contact) supplies the same voltage to contact 5 on the switch. With the motor stopped, voltage is supplied through contacts 4 and 5 of the motor centrifugal switch and the no-tumble switch in the start winding circuit. Once the motor reaches operating speed, contacts 4 and 5 of the motor centrifugal switch open and contacts 3 and 5 close. Voltage is then supplied through contact 2 of the timer and contacts 3 to 5 of the motor centrifugal switch.

From timer-contact 2, a jumper connects the voltage to timer-contact 3. Contacts 3, 4, and 5 control the buzzer and the timer motor. Contacts 3 and 4 are closed for a short period at the end of each cycle to operate the buzzer. Contacts 3 and 5 are closed whenever the timer dial is moved from the *off* position. During timed-dry operation these contacts remain closed until the end of the drying period. If either automatic cycle is selected, contacts 3 and 5 are closed for the first few minutes of the drying cycle. This allows the timer motor to run while the heating element raises the drum temperature to the drying level.

Timer-contact 6 furnishes current to the heating element through the heat-selector switch. In the *low* position, contacts 1 and 3 are closed, routing power through the low-heat thermostat, the safety thermostat, the heating element, and a set of contacts in the motor centrifugal switch. High heat is selected when contacts 1 and 2 are closed. Current to the heating element is supplied through the control thermostat. In timed-dry operations the thermostats turn the element on and off as necessary to maintain drying temperature. If *fluff dry* is selected, contacts 2 and 3 are open and the dryer runs in a timed cycle with no heat applied.

Opening and closing of timer-contact 6 is determined by the cycle of operation. This contact closes when the timer dial is moved from the *off* position and opens about 5 minutes before the end of the timed-dry or regular fabric cycle. It also opens about 3 minutes before the end of the permanent-press cycle.

Automatic operation depends on the cycling of the control (high-temperature) thermostat. When an automatic drying cycle is selected, timer-contacts 3 and 5 close at the start of the cycle but open after a short period. This removes current from the timer motor. The timer stops—tumbling and heat application continues. When the drum temperature trips the control thermostat, power is removed from the heating element and applied to the timer motor. As the timer motor runs, the drum temperature starts to fall. When a specified limit is reached, the control thermostat will close contacts 1 and 3 and open contacts 1 and 2. Heat is again applied while the timer stands still. This off again, on again operation continues until the timer advances to the *off* position.

The no-tumble feature is achieved with a reversing switch in the motor-start winding circuit just as it was with the gas dryer. Dryers with these cycles are equipped with a reversible motor and a special clutch allowing the motor to rotate backwards while the drum remains stationary. (For a complete discussion of that cycle, refer to the section in Chapter 8 dealing with no-tumble operation.)

SPECIAL FEATURES

Manufacturers have gone to great lengths in recent years to improve the automated features on their deluxe models. These unique features vary from model to model. They effect the overall operation so little that only the two most popular features are discussed here.

Antiwrinkle Cycle

At least two popular makes of automatic dryers feature a unique cycle that goes a long way in achieving a truly automatic wrinkle-free drying cycle. A special timer used in these machines allows the load to be tumbled periodically for short durations at the end of the automatic dry cycle. The diagram of the control circuit for this feature is shown in Fig. 9-7.

Contacts 1 and 2 are the usual on/off contacts of the timer. Through these contacts current is furnished to the timer and main drive motor. The antiwrinkle feature adds contacts 3, 4, 5, and 6 on the regular timer and a second timer motor (contacts A1 and A2). When the timer is moved from the *off* position, contacts 1 and 2 close, completing the circuit to the controls. At the end of the automatic dry cycle, contacts 5 and 6 open and contacts 3 and 4 close. With contacts 3 and 4 closed, the motor for the antiwrinkle cycle is energized. But with contacts 5 and 6 open, current is prevented from reaching either the drive motor or timer motor.

The antiwrinkle timer operates on a 5-minute cycle, closing contacts A1 and A2 for about 10 seconds at 5-minute intervals. During the 10-second period both the main drive motor and the regular timer motor are allowed to run. This tumbles the clothing for 10 seconds then stops for 5 minutes.

The main timer will hold contacts 3 and 4 closed for about 5 minutes of its cycle following the end of an automatic dry cycle. However, since the timer motor is energized only 10 seconds every 5 minutes, it will take the main timer motor approximately 2 hours to advance the 5 minutes at the end of the cycle. This means that every 5 minutes (for 2 1/2 hours) after the automatic dry cycle has been completed, the load will be tumbled for 10 seconds.

This additional tumbling will prevent wrinkles from setting in when the operator is not available to remove the clothing at the end of the cycle. When the main timer finally advances 5 minutes, contacts 1 and 2 and contacts 3 and 4 are opened and the cycle is complete.

Electronic Controls

Some dryers use an electronic circuit to control the run time of the machine. This replaces the conventional timer. In these machines, automatic drying time is controlled by the moisture content in the clothes instead of drum temperature.

Fig. 9-7. A simplified diagram of an antiwrinkle control circuit.

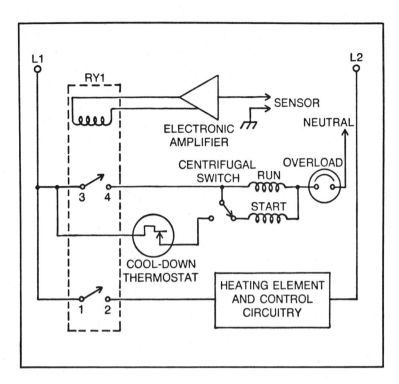

Fig. 9-8. Electronic circuits are sometimes used to control the dryer's operating cycles.

Resistance of any material is partly determined by its moisture content. Sensors placed in the drum measure the electrical resistance of the clothing. When the moisture content drops to a prescribed level, the electronic control will cause a relay to open the heating circuit and start the cool-down phase of the cycle, (Fig. 9-8).

Sensors placed in the drum make contact with damp clothing, changing the input resistance to the electronic amplifier. When this resistance indicates that further drying is needed, a relay (RY1) is energized and closes the circuit to the heating element and main drive motor. When the drum temperature reaches 120 degrees Fahrenheit, the cool-down thermostat closes its contacts. These contacts will remain closed during heat application.

When the resistance of the clothing indicates that they are dry, the electronic amplifier de-energizes the relay and opens the heating and motor circuits. Notice that the contacts of the cool-down thermostat bypass contacts 3 and 4 of the relay and keep power applied to the motor until the drum cools to 120 degrees. This added tumbling helps prevent wrinkles.

Automatic Dryers

T HIS CHAPTER DISCUSSES THE FUNDAMENTALS OF INSTALL-
ing and troubleshooting automatic dryers. Installation may
or may not be a part of your normal services. Many independent
shops contract the installation of machines sold by retail outlets,
however. A more important reason for studying installation methods
is that proper operation depends on correct installation. To service
these appliances you should be able to recognize installation conditions that could affect the operation of the machine.

Exact specifications are contained in the manufacturer's service literature. Always follow these instructions for best results.

INSTALLATION

Many of the considerations that must be observed when installing
automatic washers apply to dryer installation as well. Convenience
of access, customer's desires, and other personal considerations
must all be taken into account. The dryer must be level—front to
back and side to side. Most models are equipped with leveling feet
for this purpose.

New machines will sometimes require structural changes or
modifications to existing electric and gas lines. These changes
should be made by licensed personnel. Some service shops require
the owners to make these changes before the service technician

starts the installation. Other shops subcontract the plumbing and electrical work, requiring the owner to deal only with the service organization.

Licensing

Most local codes require that all plumbing and electrical modifications be done by persons licensed for that type of work. Your city or county engineer can give you the details on these rules; but, in general, as a service technician your work is limited to the appliance itself. Some shops work out advance agreements with local electrical and plumbing firms to do the necessary structural modifications. This is probably the most satisfactory solution for the small service shop.

All dryer installations must be vented to the outside atmosphere. Some cities insist that vent work requiring wall penetrations be done by licensed servicemen. Familiarity with the codes in your area of operation is the only way to avoid trouble. Know the requirements and limitations.

Electrical Installation

You will be concerned with two entirely different types of electrical installations. Most electric dryers require a special 240 V circuit, while the gas dryer requires 120 V for motor and control circuit operation. The requirements given here for both installations apply to most automatic dryers. You should consult the manufacturer's instructions for specific details.

Most gas-operated dryers come equipped with a heavy-duty electrical cord and require only a proper service outlet and equipment ground to complete the electrical installation. This service outlet should be capable of supplying 120 Vac and should be protected with a 20 A circuit breaker or a 15 A time-delay fuse. If at all possible a separate service outlet, used by the dryer only, should be provided.

Gas dryers are equipped with a three-prong grounding plug on the power cord. This must be plugged into a properly grounded three-prong receptacle for protection against electrical shock. Where a two-prong receptacle is encountered, it is recommended that a new service receptacle be installed by qualified and licensed electricians. Where changing the receptacle is not possible and where local codes permit, an adapter may be used to connect the three-prong appliance plug to the two-prong receptacle. Always make sure the receptacle is grounded. If not, a separate 18-gauge ground wire will have to be run from the wallplate mounting screw to a grounded metal cold water pipe as shown in Fig. 10-1. If neces-

GROUND ASSEMBLY
(ATTACH TO GROUNDED
COLD WATER PIPE)

ELECTRICALLY GROUND-
ED COLD WATER PIPE
(REMOVE PAINT, ETC.)

GROUND CLAMP
MUST BE TIGHT
ON PIPE.

GROUNDING
EYELET

GROUND WIRE (NOR18 MINIMUM)
SCREW

3-PRONG
GROUNDING PLUG

POWER SUPPLY CORD

ADAPTER
ASSEMBLY

WALL RECEPTACLE

Fig. 10-1. Using the three prong adapter.

sary, add a jumper around the water meter or any other insulated pipe connections.

Although not required by some codes, it is a good safety practice to run a separate equipment ground from any major appliance to a good ground, such as a cold water pipe. I recommend this added safety on all electrical appliances that are not frequently moved. Although three-wire heavy-duty extension cords are permitted on gas dryer installations by some local codes, the best solution is to avoid the use of any extension cord. After all, the dryer is more or less a permanent installation. Why not do it right to begin with and avoid future problems?

Requirements for an electric dryer are somewhat more complex than those of gas models. A separate 240 Vac circuit is required. Both legs of the circuit should be protected with a 30 A fuse or circuit breaker. The neutral line should remain unfused. Extension cords should never be used with 240 V dryers. In most cases, the power cord for the dryer will not be attached to a new machine. Always follow the manufacturer's instructions when installing these cords. Again, all local codes must be followed on these installations. The procedures explained here are only general guidelines.

Most local codes will permit the separate dryer circuit to terminate at a polarized three-wire wall receptacle and then connect to the dryer with a plug and cord set as shown in Fig. 10-2. This circuit must be equipped with a disconnect or pullout switch and protected with two 30 A fuses. Some local codes specify a separate circuit leading directly from the main entrance panel to the dryer. If the dryer circuit is not equipped with a disconnect or pull-

SUGGESTED DRYER INSTALLATION
(SUBJECT TO LOCAL REGULATIONS)

POWER SUPPLY FROM
MAIN ENTRANCE PANEL
120/240 V
60 Hz

DISCONNECT OR
PULLOUT SWITCH
FOR DRYER CIRCUIT

30 AMP FUSES OR
CIRCUIT BREAKER

METALLIC OR NONMETALLIC
SHEATHED CABLE

NEUTRAL WIRE
DO NOT FUSE

POLARIZED 3-WIRE
WALL RECEPTACLE
(NEMA CONFIGURATION
10-30R)

PIGTAIL TO DRYER
TERMINAL BLOCK

BLACK

WHITE

RED

TERMINAL BLOCK
ON DRYER

Fig. 10-2. A polarized three-wire 240 V connection.

out switch at the main entrance, one must be installed in the circuit between the main entrance panel and the dryer as shown in Fig. 10-3.

Local codes also vary on grounding requirements. Some communities do not allow the neutral leg to be grounded in any way. Most electric dryers are shipped with the neutral wire grounded to the frame of the dryer. Where this is not permitted, follow the manufacturer's instructions and the local codes for proper installation. A separate equipment ground is always permitted (and recommended by the service technician). This ground should be a separate wire, not connected to any electrical circuit, running from the main frame of the machine to a cold water pipe. Equipment grounding has been discussed several times in other parts of this book and is not repeated here.

A final word of caution on 240 V dryer installations. The wires that serve this appliance should always have *three* conductors. All three are current carriers and should be of equal size and insulation. Again, consult the local codes.

DISCONNECT OR
PULLOUT SWITCH
FOR DRYER CIRCUIT

30 AMP FUSES OR
(CIRCUIT BREAKER

NEUTRAL WIRE
DO NOT FUSE

METALLIC OR NON-
METALLIC SHEATHED
CABLE

BLACK

A
WHITE
B
RED
C

TERMINAL BLOCK ON DRYER

Fig. 10-3. Direct 240 V connections require a disconnect or pull-out switch.

CONVERTING 240 V DRYERS TO 120 V OPERATION

Although this type of installation is not as popular as it once was, you will occasionally encounter a 240 V dryer that has been converted for 120 V operation. Due to differences in the wiring of the various makes, we will not attempt to give you complete details of such systems. Figure 10-4 shows the wiring of one particular model. The conversion is accomplished by connecting one leg of the internal 240 V circuit to the neutral line as shown. A separate equipment ground should be used on all such installations.

Venting

All dryer installations should be vented to expel moisture, lint, and gases from the home. Indoor venting dumps moisture and lint-laden air back into the dwelling. Besides being irritating to the nostrils, lint is very flammable. In addition, byproducts of gas-fired dryers can be very dangerous to breathe and may be explosive. Always vent dryers to the atmosphere. Some codes allow under-the-house venting, but I wouldn't make such installations. I would not want explosive gases or lint to accumulate under my own home. Therefore, I refuse jobs where such installations are required. Also, the dryer should never be exhausted into a chimney.

Figure 10-5 illustrates a good through-the-wall vent installation. Where necessary, the direction of the venting can be changed by using elbows. The length of the exhaust pipe and the number of elbows should be kept to a minimum to allow the free flow of exhausted materials. When permitted, flexible ducting may be used.

Proper venting is important to obtain maximum drying efficiency. When making a rigid duct connection, the crimped end must be installed downstream of the exhaust flow as shown in Fig. 10-6. If the rigid duct connections are made incorrectly, as shown in Fig. 10-6, lint will collect at each connection and decrease the airflow.

Gas Connections

Gas connections must also be made in accordance with existing codes. Most communities specify the types of pipe, valves, and

Fig. 10-4. Converting dryers from 240 V operation to 120 V operation requires the installation of an equipment ground. (A) shows the connection for 240 V operation, (B) shows the connection for 120 V operation.

Fig. 10-5. Through-the-wall venting is recommended.

Fig. 10-6. Vent connections should point in the direction of the exhaust flow.

other fittings that may be used. Never use any of these materials without first consulting local regulations. This is especially true of any flexible tubing.

Your main concern will be with the burner itself. Most burners are designed to operate on any of the three common gases used in the United States—natural, manufactured, and liquified petroleum. Manufacturer's instructions will detail the procedure for changing burner orifices in order to use the different gases.

Be safe. Use only approved materials and procedures. Refer gas pipe work to licensed personnel.

TROUBLESHOOTING

No troubleshooting diagrams, charts, or problem—cause tables will locate every trouble for you. Good troubleshooting practices, however, dictate a logical, step-by-step method of fault isolation. We have found that troubleshooting flow charts aid in guiding a technician through these logical steps. By following a routine check-out procedure, you ensure that no steps are overlooked and reduce the time spent in pinpointing problem areas.

There are three functional areas of operation in any dryer: heating, air circulation, and cycling. Observe the operation of the dryer through each cycle and note the functional operation in each of these areas. If the unit is operating properly in all three functions, no problem exists.

Problems associated with the heat-producing section of an automatic dryer will require one kind of inspection for gas dryers and another kind for electric dryers. If you have trouble producing the required heat with gas burners, refer to the chapter explaining gas dryer operation. Heating problems in electric dryers are best located by using an ohmmeter and the manufacturer's wiring diagram to check the heating element and its control circuitry.

Air-circulation trouble can usually be pinpointed to a plugged vent or mechanical problems associated with the blower, such as a loose belt or pulley. Start at the vent outlet outside the house and see if there is adequate airflow during normal operation. If not, re-move the duct connection at the back of the dryer and check there. This will tell you whether the air is stopped inside the dryer or inside the ducting.

Cycling problems will usually be located in the timer or thermostat circuits. Observe the operation during a typical cycle and try to locate the exact point in the cycle where the failure occurs. This will usually lead you to a particular component. Voltage and resistance readings can confirm your suspicions.

Figures 10-7 and 10-8 illustrate typical troubleshooting flow charts for automatic dryers. As seen in these examples, the troubleshooting process is essentially the same for gas and electric dryers, except for problems involving the heat-generating function.

As with any troubleshooting, the first step in isolating problems in automatic dryers is to closely observe the operating cycle and determine just where in that cycle the fault occurs. Figures 10-8 and 10-9 are used to illustrate some typical examples. Perhaps the best first check is determining if the motor is starting as shown in Fig. 10-8. If the drum is turning the motor has probably

Fig. 10-7. Troubleshooting chart for a gas-powered dryer.

192

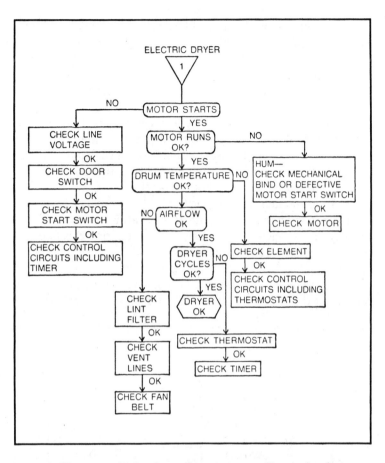

ELECTRIC DRYER

Fig. 10-8. Troubleshooting chart for an electric-powered dryer.

started. However, if the drum does not turn, listen closely to see if the motor is actually starting. Perhaps the drum is not turning because of a broken belt.

If the belt is tight and the motor is not starting, eliminate the most obvious and simplest possibilities first. For example, in Fig. 10-8 the door switch being ajar will prevent the motor from starting. The next step is to ascertain the presence of line voltage at the dryer receptacle. Once these obvious problems have been checked and the problem remains, the more difficult possibilities will have to be investigated. This will usually require the removal of one or more covers.

Notice there are usually two paths for motor current: A starting path (contacts L to 1 in the timer) to give the motor its initial starting current, and the run path (L to contact 2) for the run current. If contacts L to 1 are not made, the motor will fail to start. Although contacts L to 2 close at the same time L to 1 is closed, the centrifugal switch is in the start position and contacts 3 to 5 are open. This limits initial starting current to that supplied from contact 1.

Operation can be checked by testing for the presence of voltage at pin 1 of the timer while the start switch is held closed. It can also be checked with an ohmmeter by unplugging the dryer and connecting the meter between pin L and pin 1 of the timer. With the start switch held closed the meter should read zero ohms.

Since the overload and centrifugal switch are usually a part of the drive motor, and access to the motor is frequently restricted, these should be checked only after other possibilities have been eliminated. If voltage is present at pin 5 of the centrifugal switch in Fig. 10-9 with the start switch held closed, the motor circuits should be tested as outlined in Chapter 2.

If the motor hums when trying to start there could be a mechanical restriction on the motor shaft. If the shaft is free to turn, look for problems in the centrifugal switch, the starting capacitor, or the start winding. Frequently the switch will be stuck, due to the lint that tends to accumulate near the motor.

Once the motor is starting and running properly, the temperature within the drum should be checked. Excessively high temper-

Fig. 10-9. Electric dryer wiring diagram.

atures are usually due to a failure in the air circulating system. Look for restricted air flow paths and broken fan belts first. If all is in order in the air system, look for stuck contacts in the thermostats. This does not happen frequently, and can be very hard to detect when it does occur. The best way to check the high and low temperature thermostats is by substitution.

Low drum temperature is often caused by a clogged filter or vent duct. If the filter and ducts are clean and there is adequate air flow, do a low temperature check for voltage between L1 and L2 at the dryer receptacle. Although the motor may be running, the supply voltage can still be at fault in this situation. Only 120 Vac is required for motor operation while the heater circuit is across the 240 Vac supply.

With correct voltage present at the receptacle, the problem is in the heating element circuit. An ohmmeter can be used to check the thermostats and the element once the dryer has been unplugged. Timer and heat selector contacts are also suspect.

Most dryers have a set of safety contacts located in the motor's centrifugal switch that prevent element current until the motor is turning at the proper speed. Occasionally this switch sticks in the *open* position, thus opening the element circuit. To check operation of the switch, set a voltmeter for 240 Vac and connect it across the switch contacts. When the motor come up to operation speed the meter should read zero volts. Any reading of more than two or three volts is justification to replace the switch.

Residential Water Systems

CLEAN, CLEAR, POTABLE WATER IN ABUNDANCE IS TAKEN for granted by most homeowners. Little thought is given to the water supply—until something goes wrong. That's where you come in. As an appliance technician you probably will not be called on to do much work on water systems. In fact, many communities require that hot water and water-treatment equipment be installed and serviced by licensed plumbers. Although licensing is not a consideration in our area, my shop does not service this equipment. We feel it is better to leave this work to shops equipped to give complete service—including installation.

Why include this chapter in a book about laundry appliances? Because the water used by laundry appliances has a bearing on their efficiency. Insufficient hot water and rusty or dirty water can result in a poor wash. Chances are, you will get the call before the plumber. You need to know a little about how this equipment works in order to understand how it can affect you. When you get a complaint about poor washing, you'll be the one to determine whether the laundry appliances or water system is at fault.

WATER HEATING AND STORAGE

Although referred to as a heater, the residential water heater not only heats the water as needed but stores it for ready use. Size, shape, and capacity will vary from model to model, depending on its intended use. Heating may be either gas or electric. In either case, the water is kept at the selected temperature.

Gas-Powered Water Heaters

A typical gas-powered water heater installation is shown in Fig. 11-1. The location of this appliance should be selected on the basis of accessibility to gas and water lines, ventilation, drainage, and protection against freezing. In addition, the floor supporting the heater should be level and strong enough to support the weight of the heater when it is filled with water. No flammable materials should be stored in the vicinity of gas-powered water heaters.

Proper venting is one of the most important installation considerations. Flue gases must be removed from the dwelling and an adequate supply of fresh air must be available for combustion. The flue run should be as direct as possible with a minimum number of bends and turns. Although the flue may be vented to a chimney, it should not connect to any chimney serving an open fireplace. The heater should not be installed in a small enclosure unless one side of the enclosure is permanently open to an air supply.

Regulations regarding sizes and types of pipe for proper installation are contained in local building codes. They must be followed if you install these units. Most codes require that a combination

Fig. 11-1. A typical gas powered water heater installation.

198

RELIEF VALVE

FLUE

WATER TANK

COLD WATER

HOT WATER

INSULATION

OUTSIDE COVER

PROBE

THERMOSTAT

DRAIN

BURNER

Fig. 11-2. A cutaway view of a typical gas powered water heater.

temperature and pressure-relief valve be installed on the tank. Normally installed in the tank-opening marked *relief valve*, this device is for protection against excessive temperature and pressure in the tank. A drainpipe should lead from the relief valve to a plumbing drain or floor drain.

Figure 11-2 is a cutaway drawing that illustrates the basic operating features of a gas-powered water heater. Cold water enters the tank from the water supply system. As hot water is drawn off near the surface of the tank, additional cold water is added at the bottom of the tank. When the water temperature (as sensed by the thermostat probe) drops to a sufficiently low level, the thermostat opens the gas supply to the burner. With the burner ignited, the water temperature will start to rise. When the probe senses that the water temperature has reached the level selected on the thermostat, the burner will cut off.

A more detailed representation of the burner is shown in Fig. 11-3. In many respects, this burner operates like a gas dryer with a standing pilot. The thermostatically controlled valve has a pilot bypass feature that permits the pilot to remain lit at all times. A lever at the top of the thermostat housing provides a means for ad-

1. GASCOCK HANDLE
2. THERMOSTAT
3. PILOT TUBE
4. BURNER SUPPLY TUBE
5. AIR SHUTTER
6. MAIN BURNER
7. PILOT BURNER
8. THERMOCOUPLE
9. THERMOSTAT PROBE
10. GASLINE FITTING

justing the water temperature. When the thermostat probe senses that the water temperature is below the level selected at the thermostat, the valve opens and gas entering the burner is ignited by the pilot. This pilot has a safety feature that turns off the gas to the burner anytime the thermocouple does not sense the presence of a flame at the pilot.

Fig. 11-3. The gas burner for a water heater is thermostatically controlled.

Electric-Powered Water Heaters

Installation requirements for electric-powered water heaters vary somewhat from those for gas-powered water heaters. Venting and air circulation requirements for gas-powered water heaters are very important. Electric models do not use combustion to generate heat, so these requirements need not be considered. General construction features are the same for both heaters.

Specifications for the power connections to electric-powered water heaters will vary somewhat with local codes and the particular unit being installed. In general, however, water heaters are not connected to the neutral lines unless specifically required. A 240 V, three-wire, single-phase system connected as shown in Fig. 11-4 is typical. This should be a separate circuit protected by a circuit breaker or fuses in each leg.

Three popular types of heating units are found in electric water heaters. The simplest of these is the *single-element* heater. Operation is very similar to an electric-dryer heating element with a thermostat controlling the current flow through the heater coil as shown in Fig. 11-5(A). Figure 11-5(B) illustrates the operation of a *two-element* (duplex) water heater circuit. Two separate elements, one near the top and the other near the bottom of the tank, are controlled by separate thermostats. Each element operates independently of the other, heating the water in its particular area of the tank. Both will operate at the same time if the entire tank is cold.

The *load-limiter* water heater also has two elements but operates in a somewhat different manner than the duplex circuit. As shown in Fig. 11-5(C), the top element operates when contacts 1 and 2 of that thermostat are closed. Once the upper level has been heated, contacts 1 and 2 open, closing contacts 1 and 4. If the water in the lower portion of the tank is cold, the contacts of that thermostat will also be closed, causing the lower element to operate. Notice that the lower element cannot operate until the upper level is heated. In this circuit both elements are never on at the same time.

Fig. 11-4. Typical connections for a 240 V three-wire, single-phase, electric water heater.

Servicing

Routine servicing of water heaters is limited to occasionally

TWO "HOT" 240 V LEGS TO BREAKER BOX NO NEUTRAL IN MOST CASES

FUSED ELECTRICAL SUPPLY SWITCH

ALTERNATE LOCATION

TEMPERATURE AND PRESSURE RELIEF VALVE STEM MUST EXTEND INTO TOP 2" OF TANK

COLD WATER INLET

HOT WATER OUTLET ACCESS OPENING

ELECTRICAL LEADS

COLD WATER SHUT-OFF VALVE MUST BE IN OPEN POSITION WHILE HEATER IS IN OPERATION

UNIONS IN WATER CONNECTIONS POSITIONED ABOVE TOP LEVEL OF HEATER DIELECTRIC UNIONS RECOMMENDED IF SPECIAL DIELECTRIC NIPPLES NOT ALREADY PROVIDED BY HEATER MANUFACTURER

ACCESS PANEL

WATER CONNECTIONS IN HEATER ARE ¾" TAPERED PIPE THREAD

DRAIN VALVE USE TO PREVENT ACCUMULATION OF SEDIMENT MONTHLY, DRAIN A FEW GALLONS TO REMOVE SEDIMENT

RELIEF VALVE DISCHARGE PIPING LOWER END OF PIPE TO BE IN POSITION TO DISCHARGE INTO FLOOR DRAIN

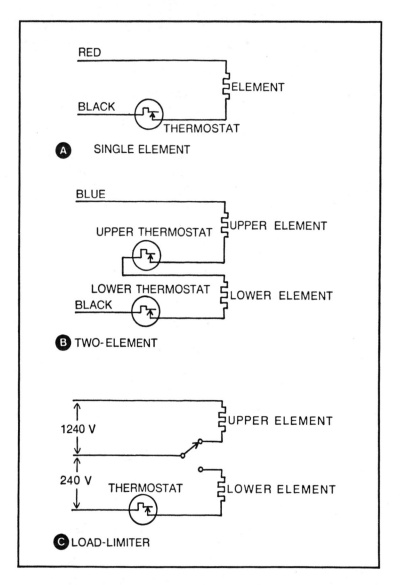

RED

ELEMENT

BLACK

THERMOSTAT

A SINGLE ELEMENT

BLUE

UPPER THERMOSTAT

UPPER ELEMENT

LOWER THERMOSTAT

BLACK

LOWER ELEMENT

B TWO-ELEMENT

UPPER ELEMENT

1240 V

240 V

THERMOSTAT

LOWER ELEMENT

C LOAD-LIMITER

draining the tank to remove sediment. Some manufacturers recommend that a few quarts of water be drawn every month to prevent accumulation of sediment in the bottom of the tank.

Troubleshooting and repairs of water heater burners and heating elements are similar to that for clothes dryers. Voltage and resistance checks will reveal most troubles in an electric water heater, while problems in a gas model can usually be located by observing its operation.

Fig. 11-5. Wiring for electric-powered water heaters, (A) single element, (B) two element, (C) load limiter.

WATER TREATMENT SYSTEMS

In localities where the residential water supply is not sufficiently pure to meet the needs of the consumers, home-water treatment systems are available to correct most deficiencies. Although the water may be potable, it can still contain minerals and other matter harmful to appliances. Deposits from the minerals contained in some water supplies can block water passages and cause corrosion of metallic parts. Water is usually analyzed and tested for four qualities—hardness, iron content, acidity, and turbidity (muddiness).

Filters

Iron, acidity, and turbidity can all be treated with filter systems. Separate filters can be used for each condition, or a complete water treatment system may be installed.

Iron suspended in the water can cause many washday complaints. In addition to its effects on water taste, iron can stain clothing and plumbing fixtures. Two common types of iron can be found in residential water systems—ferrous iron and ferric iron.

Ferrous iron (sometimes called clear-water iron) can be detected by drawing a glass of water from the faucet and allowing it to stand. At first, the water will appear to be clear. The air starts to oxidize the iron in the water. The water turns yellow and then brown, depending on the amount of ferrous iron. Although a softener can remove ferrous iron in small amounts, a separate filter system must be used to remove large quantities of this mineral.

Ferric iron (sometimes called red-water iron) is clearly visible when the water is drawn from the faucet. The water will appear yellow or reddish-brown. After standing for a period, the suspended particles will settle to the bottom of the glass. A filter system is the only answer to ferric iron problems.

Water is tested for acidity on the pH scale. A measurement less than 6.8 is considered acidic and requires a neutralizing filter to correct the situation. A filter system can also be used to remove sediments such as silt or clay (called turbid water).

The schematic diagram of a typical filter system is shown in Fig. 11-6. Untreated water is routed to the filter through the main cut off valve and the first filter isolation valve. Treated water from the filter is then routed through the output isolation valve to the various points of use.

To facilitate service of the filter without interrupting the water service, a bypass manifold is provided. During normal operation all the valves in Fig. 11-6 are in the open position with the exception of the bypass valve. The bypass valve remains closed any time the filter is in operation. When the filter is serviced, the bypass valve is opened and both isolation valves are closed. This removes the flow of water to the filter to allow it to be serviced, while at the same time routing untreated water to the service.

In some instances the bypass manifold may be replaced with

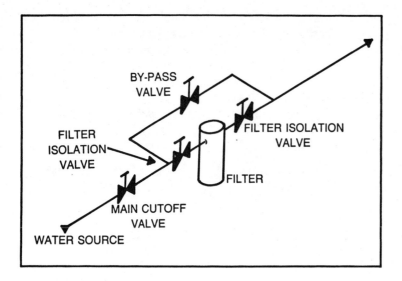

BY-PASS
VALVE

FILTER
ISOLATION
VALVE

FILTER ISOLATION
VALVE

FILTER

MAIN CUTOFF
VALVE

WATER SOURCE

a specially designed multi-position valve that provides the same function. This same type system can be used with any size filter installation.

For applications requiring little treatment and a small flow rate, a replaceable element filter similar to the one in Fig. 11-7 is frequently used. These units are about the size of a one-half gallon jar and are sometimes plumbed into the system near the water heater or under the cabinet. Several types of elements are available. The two most popular are plain and charcoal. The plain element is a porous fibrous cylinder with a hollow center. This filter will remove suspended solids and is frequently used to treat turbid water.

Where odor and taste are a problem, charcoal-impregnated elements are used. Water from the inlet pipe is routed through passages in the cap, into the bowl. This untreated water totally surrounds the element and pressure from the inlet line forces the water through the element. Acting as a screen, the element prohibits passage of most suspended particles by trapping them inside the bowl. Treated water is forced up through the center of the element to the outlet pipe. If the filter is impregnated with charcoal, the odor and taste of the water will be improved.

Trapped particles in the filter element tend to reduce water pressure at the output. When the water in bowl becomes cloudy, or whenever pressure is greatly reduced, the element should be changed. If water service must be maintained, the bypass valve should be opened and both isolation valves closed. The bowl can now be unscrewed from the cap, removing the element with the bowl. Be sure to clean the bowl each time the element is replaced.

Larger filter installations are required when the demand rate or contamination level exceeds the capabilities of the small units

Fig. 11-6. Schematic of water filter system.

just described. The larger units do the same job as the smaller ones, the difference is the size of the filter and the type of filter medium employed. Figure 11-8 illustrates the major components of the larger systems.

The tank contains a porous grout filter plate on which the filter medium rests. Composition of the filter medium will depend upon what water condition is being treated. For undissolved solids, turbidity, and similar conditions, a fine filter sand is used. Charcoal is used for odor and taste, while other chemicals may be added for other conditions.

During normal operation water enters the top of the tank through the main cut-off valve and the operating valve. The operating valve is a multi-position one, used in place of the bypass valve manifold in Fig. 11-6. In the operate position the valve routes water from the inlet into the top of the tank and from the bottom of

Fig. 11-7. The replaceable element filter.

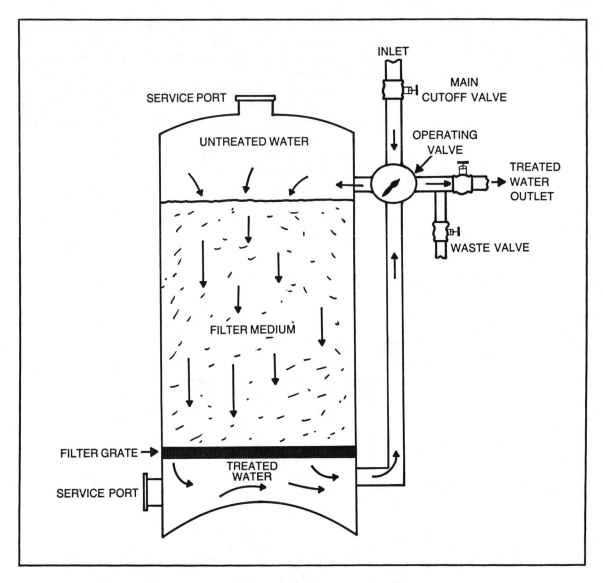

the tank to the outlet. The waste valve is closed during normal operation.

Water under pressure is forced down through the filter medium in which the suspended solids are trapped. The grate at the bottom of the filter medium is a porous substance resembling sand which has been mixed with epoxy. The grate allows the water to pass from the upper chamber into the lower chamber while restricting the flow of solids. Treated water is forced from the bottom chamber of the tank, up the connecting pipe, to the operating valves. In the operating position the valve passes the treated water to the outlet.

Fig. 11-8. A filter tank.

Most of these operating valves have a bypass position that allows water to pass directly from the inlet to the outlet without entering the tank. This position is used when it is desirable to maintain a flow of water to the outlet while the tank is being serviced.

As more and more particles become trapped in the filter medium, outlet pressure will drop and the effectiveness of the filter may be decreased. When this occurs, the filter will need to be backflushed. To backflush the system, the valve in the treated water outlet is closed. The operating valve is then placed in the backflush position and the waste valve opened. Water is passed through the filter in a reverse direction and out through the waste pipe, carrying away the trapped particles. In the backflush mode water enters the bottom of the filter medium and goes out to the operating valve. With the outlet valve closed, contaminated water is prevented from entering the outlet service pipe and is forced out through the waste valve.

Water Softeners

Water hardness is measured in mineral grains per gallon. Water softeners are designed to remove enough of these minerals to eliminate the effects of hard water. Too many minerals in the water supply reduce the detergents' ability to release their chemicals. In addition, mineral deposits in mixing valves cause premature failure of these parts.

When you encounter complaints about the lack of sudsing action obtained from detergents, look into the possibility of hard-water problems. A close check of the plumbing sometimes reveals that the softener is installed in the hot water line only. Water softeners must serve both hot and cold water lines to do their job effectively. For example, an automatic washer will use both hot and cold water in the regular cycle of routine washing. Mixing the treated hot water with the untreated cold water will reduce the overall effectiveness of the water softener.

In physical appearance softeners often resemble the larger filter systems, and are equipped with bypass manifolds or valves. Internally, however, they are quite different and will vary considerably in design. For this reason service of such units is better left to experts.

Each softener, as well as each filter, will have its peculiarities, depending on the manufacturer. Always consult the manufacturer's service manual on these devices before advising the customer. Although you may never repair these appliances, you should be aware of their effects on the water that serves the appliances you do repair.

Index

Index

Edited by Cherie R. Blazer

Other Bestsellers From TAB

☐ **ELECTRIC MOTORS— SELECTING, PROTECTING AND SERVICING—(AAVIM)**

Now with the clear, practical guidelines in this illustrated guide, you can choose the right electric motor for any job . . . and use it to cut your workload in half! All the advantages of electric motors— reasonable cost, simplicity of operation, long life, quiet, safe, compact, no exhaust fumes, and more—can be yours . . . if you have the working knowledge provided here of what electric motors are available, what they can be used for, and their limitations. 160 pp., 80 illus.
**Paper $12.95   Hard $19.95
Book No. 2663**

☐ **HOME ELECTRICAL WIRING AND MAINTENANCE MADE EASY—(AAVIM)**

With this exceptional sourcebook as your guide, you'll be amazed at how simple it is to learn professional wiring techniques— whether you want to replace a faulty switch, install a light dimmer, inspect your home's wiring system for safety hazards, or install a complete wiring system from scratch! Plus, you'll have the security and satisfaction of knowing that every procedure recommended in this wiring guide is in compliance with the most recent National Electrical Code requirements and that every possible safety precaution has been observed. 272 pp., 550 illus., most in full color. Extra Large 8″ × 11″ Format. 4-Color Throughout.
**Paper $19.95   Hard $28.95
Book No. 2673**

☐ **101 PROJECTS, PLANS AND IDEAS FOR THE HIGH-TECH HOUSEHOLD—Knott and Prochnow**

Even professional interior designers will be impressed with the expert tips on how to create a high-tech look for every room in the house. the authors show how to paint with light . . . give advice on the use and installation of track lighting . . . provide energy-saving window treatments . . . give hints on room acoustics and traffic control . . . even guidelines for selecting ergonomic seating, space-age floor coverings, ceiling materials, bed coverings, and counter top storage. There are ideas for planning a model child's room, for making better use of room space, and for giving any room a special individualized personality! 352 pp., 176 illus. Large, Easy-To-Use Format (7″ × 10″).
**Paper $16.95   Hard $24.95
Book No. 2642**

☐ **111 YARD AND GARDEN PROJECTS—from boxes and bins to tables and tools—Blandford**

Expert woodworker and metalcrafter Percy Blandford gives you step-by-step building guidance that's based on years of experience as a teacher and author of bestselling project guides. Plus, he provides a wealth of basic know-how on how to choose and work with wood and metal. He gives invaluable tips on constructing tool handles. And he shows how to custom design exactly the outdoor tools, accessories, and gardening aids that are right for your own particular needs!
**Paper $16.95   Hard $25.95
Book No. 2644**

Other Bestsellers From TAB

☐ **PLANNING AND BUILDING FENCES AND GATES—AAVIM**

This colorfully illustrated guide gives you all the expert, step-by-step guidelines and instructions you need to plan and build durable, cost-effective fences and gates. You will be able to design and construct just about any kind of fence you can think of—barbed wire, woven wire, cable wire, mesh wire, board fences, electric fences, gates, and much more! 192 pp., 356 illus. 8 1/2″ × 11″. 2-Color Throughout.

**Paper $14.95   Hard $22.95
Book No. 2643**

☐ **HOW TO CAST SMALL METAL AND RUBBER PARTS—2ND EDITION—Cannon**

Using this excellent sourcebook as a guide, you can easily make high quality, defect-free castings for almost any purpose and at an amazingly low cost! Just some of the countless uses you'll find for this potentially profitable skill . . . making obsolete or vintage car parts, hood ornaments, garden and fireplace tools, kitchen utensils, automotive parts, replacing broken antique parts, reproducing sculpture, plaques and other art . . . all kinds of decorative and useful objects for your own use or to sell! 176 pp., 143 illus.

**Paper $9.95   Hard $15.95
Book No. 2614**

☐ **HOW TO TROUBLESHOOT AND REPAIR ANY SMALL GAS ENGINE—Dempsey**

Here's a time-, money-, and aggravation-saving sourcebook that covers the full range of two and four-cycle gas engines from just about every major American manufacturer—from Briggs & Stratton, to West Bend, and others! With the expert advice and step-by-step instructions provided by master mechanic Dempsey, you'll be amazed at how easily you can solve almost any engine problem. 272 pp., 228 illus.

**Paper $10.95   Hard $21.95
Book No. 1967**

☐ **66 FAMILY HANDYMAN® WOOD PROJECTS**

Here are 66 practical, imaginative, and decorative projects . . . literally something for every home and every woodworking skill level from novice to advanced cabinetmaker: room dividers, a freestanding corner bench, china/book cabinet, coffee table, desk and storage units, a built-in sewing center, even your own Shaker furniture reproductions! 210 pp., 306 illus. 7″ × 10″.

**Paper $14.95   Hard $21.95
Book No. 2632**

☐ **UPHOLSTERY TECHNIQUES ILLUSTRATED—Gheen**

Here's an easy-to-follow, step-by-step guide to modern upholstery techniques that covers everything from stripping off old covers and padding to restoring and installing new foundations, stuffing, cushions, and covers. All the most up-to-date pro techniques are included along with lots of time- and money-saving "tricks-of-the-trade" not usually shared by professional upholsterers. 352 pp., 549 illus. 7″ × 10″.

**Paper $16.95   Hard $27.95
Book No. 2602**

☐ **THE COMPLETE BOOK OF LOCKS AND LOCKSMITHING—2nd Edition**

If you've ever spent hours trying to locate a locksmith because you locked yourself out of your home or car, here's a book that can show you how you can do all your own locksmithing chores, easily and inexpensively—even if you're an inexperienced do-it-yourselfer! If you're looking for a profitable full- or part-time business that's virtually recession proof, this book also has all the information you'll need—from locksmithing techniques to laws that apply to the trade, from the types of key blanks and machines you'll need to know how to lay out your shop and advertise your services! 352 pp., 637 illus. 7″ × 10″.

Paper $15.95   Book No. 1530

Other Bestsellers From TAB

☐ **MOPED MAINTENANCE AND REPAIR**

Guaranteed to save you time, money, and inconvenience, it is packed with practical troubleshooting and repair information, preventative maintenance tips, safety advice, even important facts on state moped regulations and operating guidelines. Generously illustrated, it provides solutions to these routine problems that plague every moped rider—flat tires, fouled spark plugs, broken or worn cables and chains, and more. 256 pp., 229 illus.
Paper $14.95   Book No. 1847

☐ **58 HOME SHELVING AND STORAGE PROJECTS—Blandford**

From a two-shelf book rack or table-top organizer to a paneled chest, basic room divider, or hall locker . . . from shelves or a spoon rack to a period reproduction of a Shaker cabinet or a Welsh dresser, you'll be amazed at the variety of projects included. And, each one includes easy-to-follow, step-by-step directions, plenty of show-how drawings, and complete materials list. 288 pp., 227 illus. 7" × 10".
Paper $14.95   Book No. 1844

☐ **HOUSE CRAFTSMANSHIP: A GUIDE TO RESTYLING AND REFURBISHING**

An information-packed sourcebook for every homeowner and do-it-yourselfer! Here are all the practical tips and step-by-step procedures you need to accomplish your restyling and refurbishing projects quickly and easily. Like having a skilled craftsman at your side. This illustrated guide shows you how to do just about every interior and exterior job imaginable, easily and inexpensively. 233 pp., 291 illus. 7" × 10".
**Paper $14.95   Hard $22.95
Book No. 1809**

☐ **WHAT'S IT WORTH? A HOME INSPECTION AND APPRAISAL MANUAL—Scaduto**

Here's a guide that can save home buyers and home owners thousands of dollars in unexpected maintenance and repair costs! You'll find out what types of structural problems occur in older and in new homes, even condominiums . . . cover everything from foundations and crawl spaces to attics and roofs . . . learn simple "tricks of the trade" for spotting problems and discover how professional appraisal techniques can be applied to any home! 256 pp., 281 illus., 7" × 10".
**Paper $12.95   Hard $21.95
Book No. 1761**